Symbiotic Interactions - From Mutualistic Alliances to Parasitic Exploits

Edited by Habib Ali

Published in London, United Kingdom

Symbiotic Interactions - From Mutualistic Alliances to Parasitic Exploits
http://dx.doi.org/10.5772/intechopen.1003506
Edited by Habib Ali

Contributors
Anna A. Kafintu-Kwashie, Bekir Bülent Arpacı, Carlos Juan Alvarado-López, Diana Ahu Prah, Doğancan Kahya, Evans Nii, Habib Ali, Hatice Nilufer Yıldız, Henrietta Mensah-Brown, José Alberto Gío-Trujillo, Pakize Gök Güler, Pinar Arıdıcı Kara, Refik Bozbuga, Selorme Adukpo, Yaw Aniweh, Ümit Haydar Erol

First published in London, United Kingdom, 2025 by IntechOpen
IntechOpen is the global imprint of INTECHOPEN LIMITED, registered in England and Wales, registration number: 11086078, 167-169 Great Portland Street, London, W1W 5PF, United Kingdom

For EU product safety concerns: IN TECH d.o.o., Prolaz Marije Krucifikse Kozulić 3, 51000 Rijeka, Croatia, info@intechopen.com or visit our website at intechopen.com.

British Library Cataloguing-in-Publication Data
A catalogue record for this book is available from the British Library

Symbiotic Interactions - From Mutualistic Alliances to Parasitic Exploits
Edited by Habib Ali
p. cm.
Print ISBN 978-0-85466-969-1
Online ISBN 978-0-85466-968-4
eBook (PDF) ISBN 978-0-85466-982-0

If disposing of this product, please recycle the paper responsibly.

Meet the editor

Dr. Habib Ali is an esteemed researcher specializing in entomology and insect pest control. With a strong academic background and a passion for scientific inquiry, he has made significant contributions to the field through his research work. Dr. Ali earned his Ph.D. in Agricultural Entomology and Insect Pest Control, providing him with a solid foundation for his future endeavors. Dr. Ali has been deeply committed to conducting impactful research and disseminating his findings to the scientific community throughout his career. His dedication is reflected in his extensive publication record, with over 150 international research papers published in renowned journals such as Frontier in Microbiology, Molecular Phylogenetic and Evolution, Insects, Journal of Economic Entomology, and Microbial Pathogenesis. Dr Ali's outstanding academic achievements and research prowess have been recognized through various awards and honors. In 2018, he was bestowed with the esteemed title of Best International Student of the Year and was the recipient of the Best Thesis Award by the China Scholarship Council (CSC) and the Plant Protection Department, FAFU, China. These accolades were a testament to his exceptional research capabilities and dedication to academic excellence. In further recognition of his contributions, Dr Ali was nominated for the World Scientist Index in 2021-24 and received the Council of Asian Science Editorship award. These accolades underscore his prominence as a leading figure in the scientific community and his commitment to advancing knowledge in his field of expertise. Dr. Ali actively participates in international conferences, symposiums, and workshops to broaden his horizons and stay at the forefront of scientific advancements. These events allow him to engage with fellow researchers, exchange ideas, and remain updated with the latest developments in his study area. Additionally, he consistently pursues educational and faculty training programs to enhance his skills and expand his intellectual capacity. Besides his research endeavors, Dr. Ali generously serves as an editor and reviewer for national and international journals, volunteering his expertise to maintain high standards in scientific publications. His multidisciplinary approach and dedication to finding sustainable solutions for managing agricultural pests contribute to advancing the field and promoting crop health. Overall, Dr. Habib Ali's remarkable research achievements and commitment to academic excellence make him a prominent figure in entomology and insect pest control.

Contents

Preface

Symbiosis, in its many forms, has long fascinated the scientific community. It provides a fundamental framework for understanding ecological interactions, evolutionary adaptations, and biological complexities. From mutually beneficial partnerships to exploitative interactions, the study of symbiotic interactions illuminates the complex relationships that shape ecosystems, influence species survival, and drive evolutionary change.

Symbiotic Interactions – From Mutualistic Alliances to Parasitic Exploits presents a comprehensive exploration of the diverse and dynamic interactions that occur across biological systems, spanning from microscopic associations to large-scale ecological frameworks.

This book explores the mechanisms, ecological significance, and evolutionary implications of symbiosis, emphasizing the roles of microbes, plants, and animals in creating complex relationships. The chapters collectively examine various symbiotic interactions, including mutualism, commensalism, parasitism, and endosymbiosis, emphasizing their functional roles in biodiversity, agriculture, and disease ecology. Special attention is given to microbial symbioses, host-microbe interactions, and the impact of symbiosis on ecosystem stability and resilience in a changing environment.

Through contributions from leading researchers, this book integrates classical knowledge with cutting-edge discoveries, offering novel insights into the molecular, genetic, and ecological underpinnings of symbiosis. By encompassing both fundamental and applied aspects, this compilation aims to serve as a valuable resource for researchers, academics, and students in the fields of microbiology, entomology, ecology, evolutionary biology, and environmental sciences.

The study of symbiotic relationships continues to push the boundaries of scientific inquiry, providing innovative strategies for sustainable agriculture, biological control, and medical advancements. Understanding the delicate balance between cooperation and exploitation within these interactions is paramount to addressing global challenges, from food security to emerging infectious diseases.

We are confident that this book will inspire further exploration and promote interdisciplinary collaborations to unlock the vast potential of symbiotic systems. We extend our sincere gratitude to all. Their dedication and expertise have made this work possible. We also acknowledge IntechOpen for providing an open-access platform to disseminate knowledge and advance scientific discourse in symbiosis research.

Habib Ali
Department of Agricultural Engineering,
Khwaja Fareed University of Engineering and Information Technology,
Rahim Yar Khan, Pakistan

Chapter 1

Introductory Chapter: Symbiotic Interactions – From Mutualistic Alliances to Parasitic Exploits

Habib Ali

1. Introduction

Symbiotic interactions are fundamental to the organization and functioning of ecosystems, shaping the evolutionary trajectories of countless species. These interactions, which occur when two or more organisms of different species live in close association, range from mutually beneficial relationships to exploitative ones. Symbiosis is a cornerstone of ecological and evolutionary biology, offering insights into the complex interdependencies that sustain life on Earth. This chapter provides an overview of the diverse forms of symbiotic interactions, from mutualistic alliances to parasitic exploits, and highlights their ecological and evolutionary significance.

2. The spectrum of symbiotic relationships

Symbiotic interactions can be broadly categorized into three main types: mutualism, commensalism, and parasitism. These categories are defined by the nature of the benefits and costs experienced by the interacting species. Mutualism involves both species deriving benefits from the relationship, commensalism involves one species benefiting, while the other is unaffected, and parasitism involves one species benefiting at the expense of the other. However, these categories are not always distinct, and many symbiotic relationships exist along a continuum, with the balance of benefits and costs shifting depending on environmental conditions and evolutionary pressures.

3. Mutualism: A win-win interaction

Mutualistic relationships are among the most fascinating and ecologically significant forms of symbiosis. In these interactions, both species involved derive clear benefits, often leading to co-evolutionary adaptations that enhance the partnership. A classic example of mutualism is the relationship between flowering plants and their pollinators. Bees, butterflies, and other pollinators obtain nectar and pollen as food sources, while plants benefit from the transfer of pollen between flowers, facilitating reproduction [1]. This mutualistic alliance has driven the evolution of specialized structures in both plants (e.g., brightly colored flowers, nectar guides) and pollinators (e.g., long proboscises, specialized pollen-carrying structures).

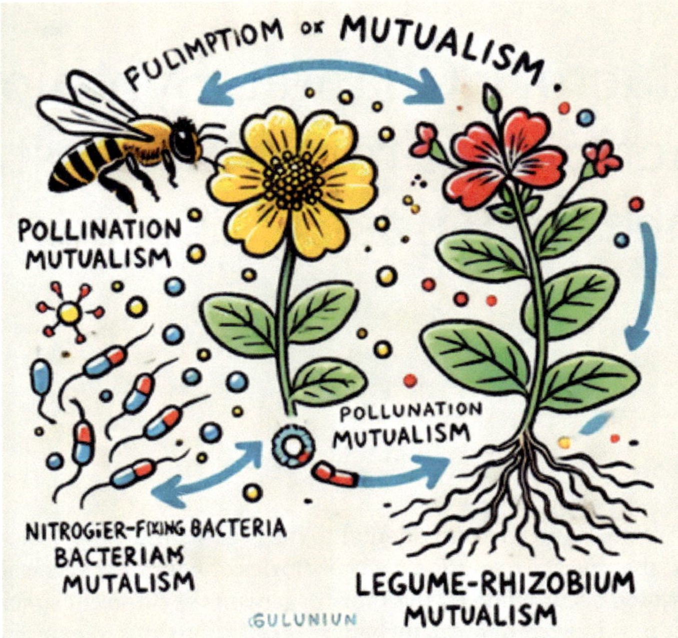

Figure 1.
Illustration of mutualism, highlighting two examples: [3] pollination mutualism, where bees collect nectar from flowers and aid in pollination, benefiting both the insect and the plant. [4] Legume-Rhizobium mutualism, where nitrogen-fixing Rhizobium *bacteria colonize legume roots, providing essential nitrogen to the plant in exchange for carbohydrates and a protective environment.*

Another well-studied example of mutualism is the relationship between legumes and nitrogen-fixing bacteria, such as *Rhizobium* [2]. These bacteria colonize the roots of leguminous plants, forming nodules where they convert atmospheric nitrogen into a form that the plant can use. In return, the plant provides the bacteria with carbohydrates and a protected environment. This mutualistic interaction is critical for soil fertility and has significant implications for agriculture, as it reduces the need for synthetic nitrogen fertilizers (**Figure 1**).

4. Commensalism: A one-sided benefit

Commensal relationships are characterized by one species benefiting, while the other is neither helped nor harmed. These interactions are often overlooked due to their seemingly neutral impact on one partner, but they can play important roles in ecosystems. For example, epiphytic plants, such as certain orchids and bromeliads, grow on the branches of trees without deriving nutrients from their hosts. They benefit from increased access to sunlight and reduced competition for resources on the forest floor, while the host tree remains unaffected (**Figure 2**).

Another example of commensalism is the relationship between remora fish and sharks. Remoras attach themselves to sharks using a specialized suction disk, gaining access to food scraps and protection from predators. The shark, however, does not appear to be significantly affected by the presence of the remora [5]. While commensal relationships may seem less dramatic than mutualistic or parasitic ones, they contribute to the complexity and stability of ecological networks.

Figure 2.
A simple scientific illustration of commensalism in insects, depicting a pseudoscorpion hitching a ride on a beetle (phoresy). The pseudoscorpion benefits from transportation to new habitats without harming the beetle.

5. Parasitism: Exploitation in nature

Parasitic interactions represent the darker side of symbiosis, where one species benefits at the expense of the other. Parasites can have profound impacts on their hosts, ranging from mild discomfort to severe debilitation or even death. Parasitism is a widespread and diverse phenomenon, encompassing a wide range of organisms, including viruses, bacteria, fungi, plants, and animals. One of the most well-known examples of parasitism is the relationship between the malaria parasite (*Plasmodium*) and humans. The parasite is transmitted to humans through the bite of an infected mosquito and invades red blood cells, causing the symptoms of malaria. This interaction has had significant impacts on human health and history, driving the evolution of genetic adaptations, such as sickle cell anemia, in populations exposed to malaria. Another example of parasitism is the relationship between mistletoe and its host trees. Mistletoe is a hemiparasitic plant that derives water and nutrients from its host while also photosynthesizing. In some cases, heavy infestations of mistletoe can weaken or even kill the host tree. Despite their negative impacts, parasites play important roles in regulating population dynamics and maintaining biodiversity [3, 4, 6].

6. Ecological and evolutionary implications of symbiosis

Symbiotic interactions have far-reaching ecological and evolutionary implications. They influence the structure and function of ecosystems, drive co-evolutionary processes, and contribute to the diversification of life. Mutualistic interactions, for example, can enhance ecosystem productivity and stability by facilitating resource exchange and promoting species coexistence. Pollination mutualism is essential for the reproduction of many plants, including crops that are critical for human food security [7].

Parasitic interactions, on the other hand, can regulate population sizes and influence community composition. By reducing the fitness of their hosts, parasites can prevent dominant species from monopolizing resources, thereby promoting biodiversity. Additionally, the arms race between hosts and parasites drives rapid evolutionary change, leading to the emergence of novel adaptations and genetic diversity. Symbiosis also plays a key role in the evolution of complex life. The endosymbiotic theory, for example, proposes that eukaryotic cells evolved from a symbiotic relationship between prokaryotic organisms. According to this theory, mitochondria and chloroplasts originated as free-living bacteria that were engulfed by a host cell, eventually becoming integrated as organelles. This evolutionary innovation laid the foundation for the diversity of complex life forms that exist today.

7. Conclusion

Symbiotic interactions, ranging from mutualistic alliances to parasitic exploits, are integral to the fabric of life. They shape ecosystems, drive evolutionary processes, and contribute to the complexity and resilience of biological systems. Understanding these interactions is crucial for addressing global challenges, such as biodiversity loss, climate change, and emerging infectious diseases. As we continue to explore the intricate web of symbiotic relationships, we gain deeper insights into the interconnectedness of life and the forces that sustain it.

Author details

Habib Ali
Department of Agricultural Engineering, Khwaja Fareed University of Engineering and Information Technology, Rahim Yar Khan, Pakistan

*Address all correspondence to: habib_ali1417@yahoo.com

References

[1] Barrett SC. Mating strategies in flowering plants: The outcrossing–selfing paradigm and beyond. Philosophical Transactions of the Royal Society of London. Series B: Biological Sciences. 2003;**358**(1434):991-1004

[2] Crespi M, Gálvez S. Molecular mechanisms in root nodule development. Journal of Plant Growth Regulation. 2000;**19**:155-166

[3] Ali H, Hou Y, Tahir MB, editors. Climate Change and Insect Biodiversity: Challenges and Implications. 1st ed. CRC Press; 2023a. DOI: 10.1201/9781003382089

[4] Ali H, Shah AN, Tahir MB, Fiaz S, Ali B. Insecticides - Advances in Insect Control and Sustainable Pest Management. IntechOpen; 2023b. DOI: 10.5772/intechopen.104285 ISBN: 978-1-80356-420-3

[5] Combes C. Parasitism: The Ecology and Evolution of Intimate Interactions. USA: University of Chicago Press; 2001

[6] Douglas AE. The Symbiotic Habit. USA: Princeton University Press; 2021

[7] Poulin R. Evolutionary Ecology of Parasites. USA: Princeton University Press; 2011

Chapter 2

Enhancing Nutritional and Sensory Quality of Plant-Based Foods through Arbuscular Mycorrhizal Fungi

José Alberto Gío-Trujillo and Carlos Juan Alvarado-López

Abstract

Arbuscular mycorrhizal fungi (AMF) play an important role in the nutritional dynamics of plants. Influencing the extraction, translocation and distribution of minerals from the soil to the plant. This has a positive effect on the mineral and nutritional content of agricultural crops. Therefore, HMA is considered a sustainable alternative, reducing the use of fertilizers by up to 50–70%. It should also be taken into account as a microbial soil resource that can contribute to self-sufficiency, food security and sovereignty in marginalized soils and areas vulnerable to population malnutrition. This chapter mentions relevant studies in relation to mycorrhizal interaction with the quality of plant foods. Recent findings have reported an increase in zinc and iron content in basic grains. Some alternative crops such as fruit, vegetables and edible foliage, have increased their phytochemical content (organic acids, phenols, carotenoids, etc.) and post-harvest properties such as texture, colorimetry and sugar content. This directly influences the taste, smell and palatability of food. In conclusion, AMF is considered a strategy to optimize and transform agro-food systems, oriented toward nutritional and sensory enrichment of plant tissues that can be used for human consumption.

Keywords: mycorrhizae, mycorrhizal symbiosis, agricultural crops, biofertilizer, mineral nutrients, vitamins, food insufficiency, food safety

1. Introduction

Global food insecurity affects more than 700 million people in the world. This means that one-third of the population does not have access to food with high nutritional value [1, 2]. Increasing socio-economic deficit and the increase in major risk factors (socio-economic well-being and health) in vulnerable areas of the population (**Figure 1**) [3].

In this context, soil, considered the core of the global food system, is the main determinant of food productivity, accessibility and nutritional quality, constituting the origin of about 95% of food [4]. Nutrient depletion in soil (marginal soils) and

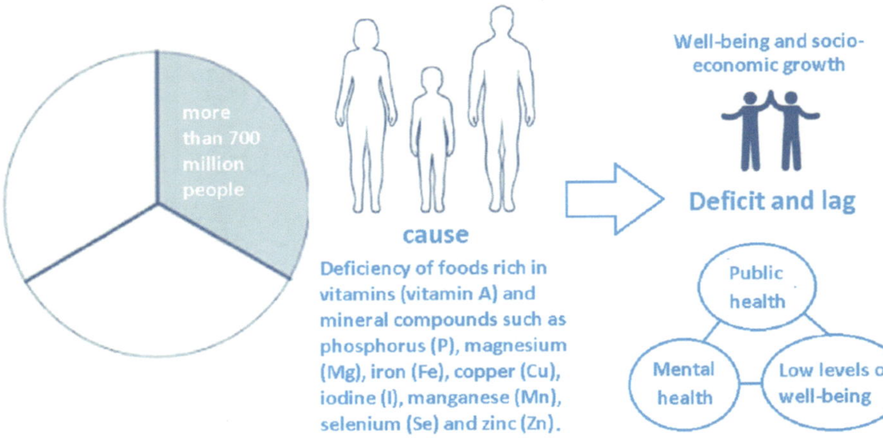

Figure 1.
Representation of the current problem of global nutrition and its main effects on socio-economic well-being (Source: 1.3).

prevailing conventional agricultural strategies such as excessive inorganic fertilization, extensive agriculture, and use of agricultural pesticides, directly influence mineral bioavailability (macro and microelements) of soils and their dynamics with agro-food systems [4, 5]. This compromises the quality of agricultural crops and their nutritional value by reducing the bioavailability of mineral elements in soil and the ability of plants to absorb, transfer and accumulate nutrients in plant tissues (roots, leaves, stems, fruits and seeds) for human consumption, which are essential to the diet and nutritional health of the world population [6, 7].

For the above, the association of beneficial microbial communities in soil is considered a novel strategy to maintain soil health, producing nutritious, safe (safe) and accessible food for the entire world population [4, 8–10]. The use of shrubby mycorrhizal fungi (AMH), considered an important microbial fraction of soils, plays a key role in modern agriculture as a promising method to increase soil productivity, profitability and fertility. Also, increasing the quality of plant-based foods [11, 12]. By intervening directly in the

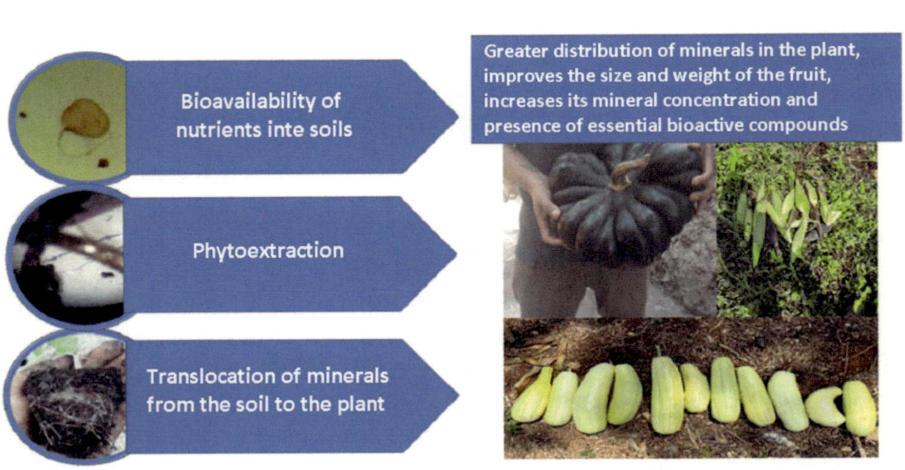

Figure 2.
Positive effects of arbuscular mycorrhizal symbiosis in relation to the quality of agricultural crops.

mineral nutrition of plants. This has a positive effect on its vigor, its efficiency, its physiology and greater transport of minerals from the soil to the plant, facilitating their accumulation and distribution in the plant biomass of associated crops (**Figure 2**) [13, 14].

2. Agricultural crops-soil microbial consortia

The efficient use of chemical fertilizers is undoubtedly the main problem of crops in modern agriculture because they oversaturate the soil with minerals, and their use does not guarantee adequate absorption by plants and efficient distribution to human-edible plant tissues [15]. Due to its efficient management in mineral nutrition of crops, directly related to crop growth, development and productivity, the association of rhizophilous microorganisms is considered a potential agricultural tool and an important soil resource for plant communities, which influences sustainable food production from plant sources [10, 16]. Therefore, the association with soil microorganisms can be considered as promising biostimulants of plant nutrition through its different mechanisms of action with host crops, with the expectation of improving nutritional and phytochemical composition of food [17, 18].

A synergistic effect has been documented between the dual inoculation of plant growth-promoting rhizobacteria (PGPR) + arbuscular mycorrhizal fungi (AMF), enhancing the positive effects of microbial interaction with host cultures [19]. Today, the expectations of the use of microbial consortia in modern agriculture are oriented toward the benefits they bring in the quality of agricultural crops and the production of food with higher nutritional contributions [20]. In this sense, dual inoculation between HMA + PGPR has reported a higher mineral content in the production of basic crops. The AMF + *Bacillus subtilis* increased the mineral profile in wheat grain and corn, mainly in the concentrations of phosphorus, nitrogen, iron and zinc [21, 22]. Also, the co-inoculation of bacteria solubilizers of microelements-HMA in soil (Selenobacteria, Zn solubilizing bacteria, among others), induces positive effects on cereal production, mainly in the increase of the concentration of Se in wheat [23] and the content of Zn in maize grains, constituting an increase of up to 10% [24].

In horticultural and fruit production, the same findings have been reported regarding the nutritional quality of the crop. In fruits of *Solanum lycopersicum*, increases in vitamin C value, organic acids and sugar content [25, 26] and an increase in phosphorus, calcium and vitamin C composition [27] were observed when using a consortium of AMF and PGPR strains (*Pseudomonas putida, Azotobacter chroococcum* and *Azosprillum lipoferum*). In other horticultural crops. In *Capsicum annuum* fruits associated with AMF + bacterial consortium of *Bacillus megaterium Pseudomonas fluorecens, Azospirillum brasilense* and *Azotobacter chrocococum*, higher concentrations of lipids and proteins were determined [28]. These benefits respond to the synergistic effects between AMF and bacterial communities in the rhizosphere, by stimulating plant metabolism [29] and promoting their mineral nutrition, producing organic acids that have a plant-based biostimulant/adjuvant effect (solubilization) of soil minerals, allowing the plant to make better use of nutrients [19, 20].

2.1 Mycorrhizal fungi

The symbiotic interaction between the mycelium of a wide variety of bound biotroph fungi belonging to the Zygomycetes class and the root system of terrestrial plants is called mycorrhiza. Glomerular mycorrhizal fungi are distributed in almost all terrestrial ecosystems, from agroecosystems to family gardens and the diversity of natural ecosystems (succession systems, low, medium, high forests, temperate forests, coastal dunes) from a wide variety of regions of the world [30]. The most important subgroup of mycorrhizal fungi is the endomycorrhizal fungi called arbuscular mycorrhizal fungi (AMH), of the phylum Glomerycota, class Glomeromycetes, which are divided into four orders and 11 families. The HMAs colonize approximately 200,000 species and about 92% of terrestrial plant families, mainly of agricultural interest such as herbaceous plants, solanaceas, grasses, legumes and tree species of interest for fruit and forest, developing different degrees of dependence [31–33].

2.2 Mycorrhizal fungi: Mineral transporters

Arbuscular mycorrhizal fungi are considered one of the main microbial components involved in soil mineral homeostasis, playing a key role as an interface between host plants and the dynamics and solubility of nutrients such as phosphorus, nitrogen, iron and zinc [34–36]. In this context, the involvement of different regulatory mechanisms and strategies for homeostasis of low-mobile minerals in soil has been documented, mainly through processes at the protein level (protein-level chelation) using glutathione, Phyto-chelatins, metallothionenes and metallochaperones. Heavy metal chelation may be favored by the presence of glomeralin, excreted by some genera of HMA, mainly those belonging to the family Glomeraceae, which stabilize the presence of minerals in the soil [12, 34].

During the symbiotic phase between the plant and a mycorrhizal arbuscular fungus, an important bidirectional flow of nutrients develops between the fungus and the plant, regulated by a mechanism of genomic transcription and activation-expression of carrier genes that enable the plant to acquire minerals that are poorly soluble or insoluble in soil [37]. For example, phosphorus, the main mineral mobilized by symbiosis, is transferred in the form of phosphates (Pi) by high-affinity transporters. Nitrogen is in turn transferred by ammonium or nitrate transporters [33]. The presence of major transcriptional activity of mineral transporters between extraradial mycelium and the cell wall of plant radical hairs (ranging from epidermal cells, cortical layer and pericycle) has been identified, through the plasma membrane of the extensive mycelial network formed by the fungus-root interaction of the plant as a first phase, followed by mobility via intracellular structures (intraradical hyphae, arbusculus, periarbuscular membrane), until the transfer of minerals to [12, 34].

Recently, a variety of high-affinity fungus-plant carrier genes have been identified as regulating the absorption of minerals absorbed from the soil into the plant [37, 38]. Mycorrhizal colonization (mycelial network) has been reported to have a positive effect on the absorption of Fe in soil, through expression of the genes HaFRO1, HaNramp1 and HaIRT1 [39]. In addition, they have been reported in the expression of other iron reductase groups (permeases) and ferroxidases such as NRAMP, RISMF1, RISMF2, RISMF3.1, RISMF3.2 [40, 41], SbDMAS2, SbNAS2 and SbYS1 [42]. As well as ZmNAS1, ZmNAS3 and ZmYS1, which are involved in the solubility and absorption of Zn/Fe in plants [43]. About 30 mineral-carrying genes such as Fe, Zn, Cu and Mn have been detected in Rhizophagus irregularis. Of these, eight main families of mineral-carrying genes have been identified, such as CTR (Cu), P1b-ATPase (Cu), SIT (SIT-Fe), OFET (Fe), VIT (Fe/Mn), ZIP (Zn, Mn), Cation diffusion facilitator (CDF) (Fe, Zn, Mn) and NRAMP (Fe, Mn, Fe/Mn) [34, 43]. In other studies, the main groups of participants are mentioned: Mn-Fe (SMF1), Cu fungal (CTR), Cu vegetal (COPt), Cu-ATPase (CCC2), Fe-Mn (CCC1.3), permeasa Fe (FTR1), Zn (ZRT1), permeasa Zn-Fe (ZIP, some also carry Mn) and CDF (Cation diffusion facilitator) [12]. Based on the above, these mechanisms regulated by an intermediate such as HMA (mycorrhizal pathway), potentially influence the processes of phyto-extraction-translocation of minerals, accumulating a higher concentration of nutrients in plants, Redistributing them more efficiently to plant tissues and biomass. This impliesn better distribution and concentration of mineral nutrients in the biomass and plant tissues such as fruits, stems, foliage, sprouts and seeds [12, 44, 45].

2.3 Mycorrhizal fungi-food security

Inoculation of agricultural crops with shrubbery fungi has been mentioned as a viable strategy to address the problem of food insecurity and insufficiency, transforming agro-food systems toward sustainable agricultural practices [11]. The Commission's proposal for a Council decision on the European Union's action program in the field of education and training. Mycorrhizal fungi are directly involved in the mineral nutrition of associated crops, playing an important role as plant biofertilizer or biostimulant [18]. Mainly by the release and stimulation of estrigolactones (SL), a phytohormone, key to greater flow in the transport, distribution and balance of minerals in plants. Reflecting a positive effect on the growth and morphological development of its host [36]. This synergy between HMA and SL increases various agronomic aspects in their phytosanitary status (increased resistance to adverse conditions, pests and diseases), vigor and agricultural productivity [46]. Over the last decade, there has been evidence of HMA's role as quality regulators in food production, focusing on increasing mineral concentrations of Zn, Fe, Cu and Mn in plant

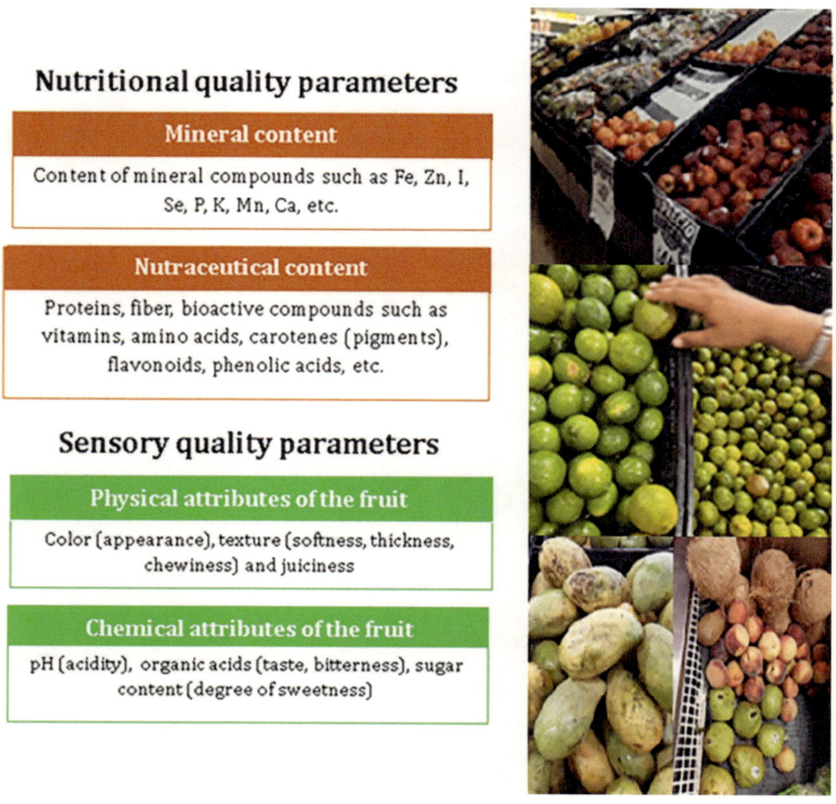

Figure 3.
Main effects of mycorrhizal interaction in relation to the quality of agricultural crops.

tissues fit for human consumption [13, 14]. Recent findings report a positive influence on the quality of agricultural grains in basic crops [47], increasing the mineral profile of N, P, Zn and Fe in cereals such as sorghum [48], chickpea [49], rice [50], barley [51], wheat [52–54] and maize [24].

In particular, the prospects of association of agricultural crops with HMA have been oriented to improve crop quality for good market acceptance and added value, by increasing the main sensory attributes (taste, color, smell, texture) and nutritional in alternative crops, such as legumes, oilseeds, edible forages, fruits and vegetables (**Figure 3**) [55–57].

Table 1 shows relevant studies on the quality of agricultural crops associated with arbuscular mycorrhizal fungi, influencing their mineral composition, nutraceuticals, weight, size, sensory and organoleptic parameters.

Species	benefited plant tissue	Main effect	References
Cucumber (*Cucumis sativus*)	Fruit	Improved weight and size development in fruit; Increased composition of phenols, flavonoids, soluble solids and antioxidant capacity	[57, 58]
Black bean (*Phaseolus vulgaris*)	Legume	Increased P, K, Cu and Zn content	[59]
Zucchini (*Cucurbita pepo*)	Fruit	Improved fruit development (weight and size)	[57]
Pumpkin (*Cucurbita moschata*)	Fruit	Improved fruit development (weight and size)	[60]
Chicory (*Cichorium intybus*,)	Foliage	Increased content of Fe, Zn, fructose, inulin and carotenoid compounds	[61]
Strawberry (*Fragaria X ananassa*)	Fruit	Increased composition of phenols, flavonoids, soluble solids and Se	[62–64]
Tomato (*Solanum lycopersicum*)	Fruit	Improved fruit development (weight and size). Increased concentrations of N, P, Cu, antioxidant capacity and carotenoid content (lycopene). Improved texture, pH, soluble solids content, phenols and organic acids of the fruit	[65–69]
Pineapple (*Ananas comosus*)	Fruit	Increase in weight and size of fruit	[69]
Raspberry (*Rubus idaeus*)	Fruit	Increased content of sugars and organic acids	[70]
Garlic (*Allium sativum*)	Bulb	Increased composition of Se	[71–73]
Onion (*Allium cepa*)	Bulb	Increased composition of Se, Ca, ascorbic acid (vitamin C), thiamine (vitamin B1), antioxidant activity and organic acids	[71, 73, 74]
Wild leeks (*Allium tricocum*)	Bulb	Improved fruit development (weight and size). Increased nutraceutical values and mineral composition	[73]
Chili pepper (*Capsicum annum*)	Fruit	Increased chlorophyll, Fe, Zn, N and P in foliage. Improved fruit development (weight and size)	[75]
Chili piquin (*Capsicum annuum var. labriusculum*)	Fruit	Increase in the content of phenols, soluble solids and vitamins	[76]

Species	benefited plant tissue	Main effect	References
Papaya (*Carica papaya*)	Fruit	Increase in sugar content	[77]
Melón (*Cucumis melo*)	Fruit	Increase in pulp thickness, peel thickness (texture), sugar content, protein concentrations, polyphenols, flavonoids, carotenoids and ascorbic acid. Increase in weight and size of fruit	[78, 79]
Moringa (*Moringa oleífera*)	Foliage	Increased composition of Cu, Zn and glucosinolates in foliage	[80]
Mint *(Mentha pulegium*)	Foliage	Increase in the content of sugars, carotenoid compounds, phenols, flavonoids and organic acids	[81]
Parsley (*Petroselinum hortense*)	Foliage	Improved content of carotenoids, phenols, flavonoids and organic acids	[81]
Lettuce (*Lactuca sativa*)	Foliage	Increased content of phenols	[82]
Celery (*Apium graveolens*)	Stem and foliage	Increased content of Se	[83]
Mandarin (*Citrus reticulata*)	Fruit	Improved fruit development (weight and size). Increased sugar, vitamin C, phenol and total flavonoid content	[84]
Orange (*Citrus* × *sinensis*)	Fruit	Improved fruit development. Increased sugar, vitamin C, phenol and total flavonoid content	[84]
Alfalfa (*Medicago sativa*)	Foliage	Increased Fe, Zn, S, Ca and P concentrations	[38]

Table 1.
Review of the main contributions of the arbuscular mycorrhizal association with alternative crops in relation to their increase in phytochemistry, nutritional and sensory quality and harvest yield (size and weight).

3. Conclusion

In general, the interaction of arbuscular mycorrhizal fungi represents an important agronomic method toward sustainability and food security, which closely influences the mineral nutrition of plants, being related to the increase in the profitability of both crops (number of fruits) and the quality of the fruit harvest (polar diameter; equatorial diameter and weight). Likewise, it has been shown that the arbuscular mycorrhizal symbiosis with a great diversity of agricultural crops turns out to be an important factor in different sensory and chemical attributes of the fruit in post-harvest. For example, the positive influence on firmness (texture), colorimetry (peel and pulp), sugar content (° Brix) and the presence of organic acids that directly influence the flavor of the fruit is mentioned.

Acknowledgements

We would like to thank the institutions TECNM, Tizimín Campus, Valle de Oaxaca Campus, CIDIR, Oaxaca Unit and CIATEJ, Sureste Unit, for providing the facilities to develop this review. We would also like to thank the TECNM PhD program, Conkal Campus and CONAHCYT for the postgraduate scholarship awarded (623916).

Conflict of interest

The authors declare no conflict of interest.

Author details

José Alberto Gío-Trujillo* and Carlos Juan Alvarado-López
National Technological Institute of Mexico, Conkal Campus, Avenida Tecnológico, Yucatán, Mexico

*Address all correspondence to: albertogio88@hotmail.com

References

[1] FAO (Food and Agriculture Organization). The State of Food Security and Nutrition in the World 2023. Urbanization, Agri-Food Systems Transformation and Healthy Diets along the Rural-Urban Continuum. Roma, Italia: FAO. FAO, FIDA, OMS, PMA, UNICEF; 2023. 48 p. DOI: 10.4060/cc3017es

[2] Wakeel A, Farooq M, Bashir K, Ozturk L. Micronutrient malnutrition and biofortification: Recent advances and future perspectives. In: Mohammad A, Takehiro K, David J, Lam-Son P, Toru F, editors. Plant micronutrient use efficiency. Academic Press; 2018. pp. 225-243. DOI: 10.1016/B978-0-12-812104-7.00017-4

[3] FAO (Food and Agriculture Organization). The State of the World's Land and Water Resources for Food and Agriculture - Systems on the Edge - Synthesis Report FAO, ONU, FAOSTAT. Roma, Italia: FAO (Food and Agriculture Organization). 308 p; 2021. DOI: 10.4060/cb7654es

[4] UIP, FAO. Nutrition and Food Systems: A Handbook for Parliamentarians No 32. UIP. Roma: FAO; 2021. 72 p. DOI: 10.4060/cb2005es

[5] FAO (Food and Agriculture Organization). Soils for Nutrition: State of the Art. Roma, Italia: FAO, ONU; 2022. DOI: 10.4060/cc0900en

[6] Das S, Green A, Fan MX. Zinc deficiency in Indian soils is associated with low crop productivity and crop quality. Better Crops–South Asia. 2019;**11**:11-14. Available from: http://www.ipni.net/publication/bca.nsf/0/FB4685DE3819A56E852583CB00625D31/$FILE/BCSA%202019-1-11.pdf [Accessed: March 23, 2023]

[7] Ahmed N, Zhang B, Chachar Z, Li J, Xiao G, Wang Q, et al. Micronutrients and their effects on horticultural crop quality, productivity and sustainability. Scientia Horticulturae. 2024;**323**:112512. DOI: 10.1016/j.scienta.2023.112512

[8] Upadhayay VK, Singh AV, Khan A, Sharma A. Contemplating the role of zinc-solubilizing bacteria in crop biofortification: An approach for sustainable bioeconomy. Frontiers in Agronomy. 2022;**4**:903321. DOI: 10.3389/fagro.2022.903321

[9] Dhiman K, Sharma D, Kumari R, Tomar P. Biofortification of crops using microbes–a promising sustainable agriculture strategy. Journal of Plant Nutrition. 2023;**46**(12):2912-2935. DOI: 10.1080/01904167.2022.2160755

[10] Kaur T, Lata Rana K, Kour D, Sheikh L, Yadav N, Nath Yadav A, et al. Microbe-mediated biofortification for micronutrients: Present status and future challenges. In: Trends of Microbial Biotechnology for Sustainable Agriculture and Biomedicine Systems: Perspectives for Human Health. Amsterdam: Elsevier; 2020. pp. 1-17. DOI: 10.1016/B978-0-12-820528-0.00002-8

[11] Upadhayay VK, Singh J, Khan A, Lohani S, Singh AV. Mycorrhizal mediated micronutrients transportation in food based plants: A biofortification strategy. In: Varma A, Choudhary D, editiors. Mycorrhizosphere and Pedogenesis. Singapore: Springer; 2019. p. 1-14. DOI: 10.1007/978-981-13-6480-8_1

[12] Ferrol N, Tamayo E, Vargas P. The heavy metal paradox in arbuscular mycorrhizas: From mechanisms to

biotechnological applications. Journal of Experimental Botany. 2016;**67**(22):6253-6265. DOI: 10.1093/jxb/erw403

[13] Lehmann A, Veresoglou SD, Leifheit EF, Rillig MC. Arbuscular mycorrhizal influence on zinc nutrition in crop plants—A meta-analysis. The Soil. 2014;**69**:121-131. DOI: 10.1016/j.soilbio.2013.11.001

[14] Lehmann A, Rillig MC. Arbuscular mycorrhizal contribution to copper, manganese and iron nutrient concentrations in crops—A meta-analysis. Soil, Biology and Biochemistry. 2015;**81**:147-158. DOI: 10.1016/j.soilbio.2014.11.013

[15] Kutman UB. Mineral nutrition and crop quality. In: Rengel Z, Cakmak I, editors. Marschner's Mineral Nutrition of Plants. Academic Press; 2023. pp. 419-444. DOI: 10.1016/B978-0-12-819773-8.00020-4

[16] Kumari K, Patra A, Sahu KP, Dilawari R, Mehta S. Nutritional biofortification of crops by microbes. In: Sustainable Agriculture Reviews 60: Microbial Processes in Agriculture. Cham: Springer Nature Switzerland; 2023. pp. 269-292. DOI: 10.1007/978-3-031-24181-9_11

[17] Agnolucci M, Avio L, Palla M, Sbrana C, Turrini A, Giovannetti M. Health-promoting properties of plant products: The role of mycorrhizal fungi and associated bacteria. Agronomy. 2020;**10**:1864. DOI: 10.3390/agronomy10121864

[18] Sun W, Shahrajabian MH. The application of arbuscular mycorrhizal fungi as microbial biostimulant, sustainable approaches in modern agriculture. Plants. 2023;**12**(17):3101. DOI: 10.3390/plants12173101

[19] Yu L, Zhang H, Zhang W, Liu K, Liu M, Shao X. Cooperation between arbuscular mycorrhizal fungi and plant growth-promoting bacteria and their effects on plant growth and soil quality. Peer J. 2022;**10**:e13080. DOI: 10.7717/peerj.13080/supp-1

[20] Melini F, Melini V, Luziatelli F, Abou-Jaoudé R, Ficca AG, Ruzzi M. Effect of microbial plant biostimulants on fruit and vegetable quality: Current research lines and future perspectives. Frontiers in Plant Science. 2023;**14**:1251544. DOI: 10.3389/fpls.2023.1251544

[21] Yadav R, Ror P, Rathore P, Ramakrishna W. Bacteria from native soil in combination with arbuscular mycorrhizal fungi augment wheat yield and biofortification. Plant Physiology and Biochemistry. 2020;**150**:222-233. DOI: 10.1016/j.plaphy.2020.02.039

[22] Yadav R, Ror P, Rathore P, Kumar S, Ramakrishna W. *Bacillus subtilis* CP4, isolated from native soil in combination with arbuscular mycorrhizal fungi promotes biofortification, yield and metabolite production in wheat under field conditions. Journal of Applied Microbiology. 2021;**131**(1):339-359. DOI: 10.1111/jam.14951

[23] Durán P, Viscardi S, Acuña JJ, Cornejo P, Azcón R, Mora M. Endophytic selenobacteria and arbuscular mycorrhizal fungus for selenium biofortification and Gaeumannomyces graminis biocontrol. Journal of Soil Science and Plant Nutrition. 2018;**18**(4):1021-1035. DOI: 10.4067/S0718-95162018005002902

[24] Suganya A, Saravanan A, Baskar M, Pandiyarajan P, Kavimani R. Agronomic biofortification of maize (*Zea mays* L.) with zinc by using of graded levels of zinc in combination with zinc solubilizing bacteria and arbuscular mycorrhizal fungi. Journal

of Plant Nutrition. 2021;**44**(7):988-994. DOI: 10.1080/01904167.2020.1845383

[25] Bona E, Cantamessa S, Massa N, Manassero P, Marsano F, Copetta A, et al. Arbuscular mycorrhizal fungi and plant growth-promoting pseudomonads improve yield, quality and nutritional value of tomato: A field study. Mycorrhiza. 2017;**27**:1-11. DOI: 10.1007/s00572-016-0727-y

[26] Bona E, Todeschini V, Cantamessa S, Cesaro P, Copetta A, Lingua G, et al. Combined bacterial and mycorrhizal inocula improve tomato quality at reduced fertilization. Scientia Horticulturae. 2018;**234**:160-165. DOI: 10.1016/j.scienta.2018.02.026

[27] Ordookhani K, Moezi A, Khavazi K, Rejali F. Effect of plant growth promoting rhizobacteria and mycorrhiza on tomato fruit quality. Acta Horticulturae. 2013;**989**:91-96. DOI: 10.17660/ActaHortic.2013.989.9

[28] Gamboa-Angulo J, Ruíz-Sánchez E, Alvarado-López C, Gutiérrez-Miceli F, Ruíz-Valdiviezo VM, Medina-Dzul K. Effect of microbial biofertilizers on the agronomic characteristics of the plant and fruit quality of Xcat'ik chili (*Capsicum annuum* L.). Terra Latinoamericana. 2020;**38**:817-826. DOI: 10.28940/terra.v38i4.716

[29] Chávez-González JD, Flores-Núñez VM, Merino-Espinoza IU, Partida-Martínez LP. Desert plants, arbuscular mycorrhizal fungi and associated bacteria: Exploring the diversity and role of symbiosis under drought. Environmental Microbiology Reports. 2024;**16**(4):e13300. DOI: 10.1111/1758-2229.13300

[30] Brundrett MC, Tedersoo L. Evolutionary history of mycorrhizal symbioses and global host plant diversity. New Phytologist. 2018;**220**(4):1108-1115. DOI: 10.1111/nph.14976

[31] Garg N, Chandel S. Arbuscular mycorrhizal networks: Process and functions. In: Lichtfouse E, Hamelin M, Navarrete M, Debaeke P, editors. Sustainable Agriculture. Vol. 2. Dordrecht: Springer; 2011. pp. 907-930. DOI: 10.1007/978-94-007-0394-0_40

[32] Tedersoo L, Bahram M, Zobel M. How mycorrhizal associations drive plant population and community biology. Science. 2020;**367**(6480):eaba1223. DOI: 10.1126/science.aba1223

[33] Camarena-Gutiérrez G. Interaction between plants and arbuscular mycorrhizal fungi. Revista Chapingo Serie Ciencias Forestales y del Ambiente. 2012;**18**(3):409-421. DOI: 10.5154/r.rchscfa.2011.11.093

[34] Tamayo E, Gómez-Gallego T, Azcón-Aguilar C, Ferrol N. Genome-wide analysis of copper, iron and zinc transporters in the arbuscular mycorrhizal fungus rhizophagus irregularis. Frontiers in Plant Science. 2014;**5**:547. DOI: 10.3389/fpls.2014.00547

[35] Jacott CN, Murray JD, Ridout CJ. Ridout, trade-offs in arbuscular mycorrhizal symbiosis: Disease resistance, growth responses and perspectives for crop breeding. Agronomy. 2017;**7**(4):75. DOI: 10.3390/agronomy7040075

[36] Bhantana P, Rana MS, Sun XC, Moussa MG, Saleem MH, Syaifundin M, et al. Arbuscular mycorrhizal fungi and its major role in plant growth, zinc nutrition, phosphorous regulation and phytoremediation. Symbiosis. 2021;**84**:19-37. DOI: 10.1007/s13199-021-00756-6

[37] Rajapitamahuni S, Kang BR, Lee TK. Exploring the roles of arbuscular mycorrhizal fungi in plant–iron homeostasis. Agriculture. 2023;**13**(10):1918. DOI: 10.3390/agriculture13101918

[38] Rahman MA, Parvin M, Das U, Ela EJ, Lee SH, Lee KW, et al. Arbuscular mycorrhizal symbiosis mitigates iron (Fe)-deficiency retardation in alfalfa (*Medicago sativa* L.) through the enhancement of Fe accumulation and sulfur-assisted antioxidant defense. International Journal of Molecular Sciences. 2020;**21**(6):2219. DOI: 10.3390/ijms21062219

[39] Kabir AH, Debnath T, Das U, Prity SA, Haque A, Rahman MM, et al. Arbuscular mycorrhizal fungi alleviate Fe-deficiency symptoms in sunflower by increasing iron uptake and its availability along with antioxidant defense. Plant Physiology and Biochemistry. 2020;**150**:254-262. DOI: 10.1016/j.plaphy.2020.03.010

[40] Li M, Wang R, Tian H, Gao Y. Transcriptome responses in wheat roots to colonization by the arbuscular mycorrhizal fungus *Rhizophagus irregularis*. Mycorrhiza. 2018;**28**:747-759. DOI: 10.1007/s00572-018-0868-2

[41] López-Lorca VM, Molina-Luzón MJ, Ferrol N. Characterization of the NRAMP gene family in the arbuscular mycorrhizal fungus rhizophagus irregularis. Journal of Fungi. 2022;**8**(6):592. DOI: 10.3390/jof8060592

[42] Prity SA, Sajib SA, Das U, Rahman MM, Haider SA, Kabir AH. Arbuscular mycorrhizal fungi mitigate Fe deficiency symptoms in sorghum through phytosiderophore-mediated Fe mobilization and restoration of redox status. Protoplasma. 2020;**257**:1373-1385. DOI: 10.1007/s00709-020-01517-w

[43] Chorianopoulou SN, Saridis YI, Dimou M, Katinakis P, Bouranis DL. Arbuscular mycorrhizal symbiosis alters the expression patterns of three key iron homeostasis genes, ZmNAS1, ZmNAS3, and ZmYS1, in S deprived maize plants. Frontiers in Plant Science. 2015;**6**:257. DOI: 10.3389/fpls.2015.00257

[44] Zhou X, Li J, Tang N, Xie H, Fan X, Chen H, et al. Genome-wide analysis of nutrient signaling pathways conserved in arbuscular mycorrhizal fungi. Microorganisms. 2021;**9**(8):557. DOI: 10.3390/microorganisms9081557

[45] Tisserant E, Malbreil M, Kuo A. Genome of an arbuscular mycorrhizal fungus provides insight into the oldest plant symbiosis. National Academy of Sciences of the United States of America. 2013;**110**:20117-20122. DOI: 10.1073/pnas.1313452110

[46] Boyno G, Rezaee Danesh Y, Demir S, Teniz N, Mulet JM, Porcel R. The complex interplay between arbuscular mycorrhizal fungi and strigolactone: Mechanisms, sinergies, applications and future directions. International Journal Molecular Sciencies. 2023;**24**:16774. DOI: 10.3390/ijms242316774

[47] Zhang S, Lehmann A, Zheng W, You Z, Rillig MC. Arbuscular mycorrhizal fungi increase grain yields: A meta-analysis. New Phytologist. 2019;**222**(1):543-555. DOI: 10.1111/nph.15570

[48] Nakmee PS, Techapinyawat S, Ngamprasit S. Comparative potentials of native arbuscular mycorrhizal fungi to improve nutrient uptake and biomass of *Sorghum bicolor* Linn. Agriculture and Natural Resources. 2016;**50**:173-178. DOI: 10.1016/j.anres.2016.06.004

[49] Pellegrino E, Bedini S. Enhancing ecosystem services in sustainable

agriculture: Biofertilization and biofortification of chickpea (*Cicer arietinum* L.) by arbuscular mycorrhizal fungi. Soil Biology and Biochemistry. 2014;**68**:429-439. DOI: 10.1016/j.soilbio.2013.09.030

[50] Gontia-Mishra I, Sapre S, Tiwari S. Zinc solubilizing bacteria from the rhizosphere of rice as prospective modulator of zinc biofortification in rice. Rhizosphere. 2017;**3**:185-190. DOI: 10.1016/j.rhisph.2017.04.013

[51] Watts-Williams SJ, Nguyen TD, Kabiri S, Losic D, McLaughlin MJ. Potential of zinc-loaded graphene oxide and arbuscular mycorrhizal fungi to improve the growth and zinc nutrition of *Hordeum vulgare* and *Medicago truncatula*. Applied Soil Ecology. 2020;**150**:103464. DOI: 10.1016/j.apsoil.2019.103464

[52] Pellegrino E, Öpik M, Bonari Ercoli EL. Responses of wheat to arbuscular mycorrhizal fungi: A meta-analysis of field studies from 1975 to 2013. Soil Biology and Biochemistry. 2015;**84**:210-217. DOI: 10.1016/j.soilbio.2015.02.020

[53] Ercoli L, Schüßler A, Arduini I, Pellegrino E. Strong increase of durum wheat iron and zinc content by field-inoculation with arbuscular mycorrhizal fungi at different soil nitrogen availabilities. Plant and Soil. 2017;**419**:153-167. DOI: 10.1007/s11104-017-3319-5

[54] Ma X, Luo W, Li J, Wu F. Arbuscular mycorrhizal fungi increase both concentrations and bioavilability of Zn in wheat (*Triticum aestivum* L) grain on Zn-spiked soils. Applied Soil Ecology. 2019;**135**:91-97. DOI: 10.1016/j.apsoil.2018.11.007

[55] Rouphael Y, Franken P, Schneider C, Schwarz D, Giovannetti M, Agnolucci M, et al. Arbuscular mycorrhizal fungi act as biostimulants in horticultural crops. Scientia Horticulturae. 2015;**196**:91-108. DOI: 10.1016/j.scienta.2015.09.002

[56] Noceto AP, Bettenfeld P, Boussageon R, Hériché M, Sportes A, van Tuinen D, et al. Arbuscular mycorrhizal fungi, a key symbiosis in the development of quality traits in crop production, alone or combined with plant growth-promoting bacteria. Mycorrhiza. 2021;**31**:655-669. DOI: 10.1007/s00572-021-01054-1

[57] Alvarado-Carrillo M, Díaz-Franco A, Alejandro-Allende F. Poultry manure, arbuscular mycorrhiza and chemical fertilization reduced productivity of zucchini and cucumber. Revista Internacional de Contaminación Ambiental. 2018;**34**(2):273-279. DOI: 10.20937/RICA.2018.34.02.08

[58] López-Morales ML, Leos-Escobedo L, Alfaro-Hernández L, Morales-Morales AE. Impact of organic fertilizers associated with mycorrhizae on yield and nutraceutical quality of cucumber. Revista Mexicana De Ciencias Agrícolas. 2022;**13**(5):785-798. DOI: 10.29312/remexca.v13i5.2868

[59] Carrara JE, Reddivari L, Heller WP. Inoculation of black turtle beans (*Phaseolus vulgaris*) with mycorrhizal fungi increases the nutritional quality of seeds. Plant-Environment Interactions. 2023;**5**:1-10. DOI: 10.1002/pei3.10128

[60] Carrara JE, Heller WP. Arbuscular mycorrhizal species vary in their impact on nutrient uptake in sweet corn (*Zea mays*) and butternut squash (*Cucurbita moschata*). Frontiers in Agronomy. 2022;**4**:1040054. DOI: 10.3389/fagro.2022.1040054

[61] Pepe A, Di Baccio D, Magnani E, Giovannetti M, Sbrana C. Zinc and

iron biofortification and accumulation of health-promoting compounds in mycorrhizal *Cichorium intybus* L. Journal Soil Science Plant Nutrition. 2022;**22**:4703-4716. DOI: 10.1007/s42729-022-00953-2

[62] Cecatto AP, Ruiz FM, Calvete EO, Martínez J, Palencia P. Mycorrhizal inoculation affects the phytochemical content in strawberry fruits. Acta Scientiarum. Agronomy. 2016;**38**(2):227-237. DOI: 10.4025/actasciagron.v38i2.27932

[63] Todeschini V, Ait Lahmidi N, Mazzucco E, Marsano F, Gosetti F, Robotti E, et al. Impact of beneficial microorganisms on strawberry growth, fruit production, nutritional quality, and volatilome. Frontiers in Plant Sciencie. 2018;**9**:1611. DOI: 10.3389/fpls.2018.01611

[64] Antoniou O, Chrysargyris A, Xylia P, Tzortzakis N. Effects of selenium and/or arbuscular mycorrhizal fungal inoculation on strawberry grown in hydroponic trial. Agronomy. 2021;**11**(4):721. DOI: 10.3390/agronomy11040721

[65] Di Martino C, Fioretto A, Palmieri D, Torino V, Palumbo G. Influence of tomato plant mycorrhization on nitrogen metabolism, growth and fructification on P-limited soil. Journal of Plant Growth Regulation. 2019;**38**:1183-1195. DOI: 10.1007/s00344-019-09923-y

[66] Mohammed HA, Amao A, Halshoy HS, Hama JR, Rashid HA, Rashid HW. Soil-borne microbes, natural stimulants, and post-harvest treatments alter quality and phytochemicals of tomato fruit. International Journal of Vegetable Science. 2023;**29**(6):1-13. DOI: 10.1080/19315260.2023.2272838

[67] Mena-Echevarría A, Ramírez-Tobias HM, Méndez-Cortés H, Rojas-Velázquez ÁN, López-Palacios C, Hipólito-Piedras RP. The origin and type of inoculum determine the effect of arbuscular mycorrhizal fungi on tomato under different irrigation regimes. Agronomy. 2024;**14**:1687. DOI: 10.3390/agronomy14081687

[68] Pellegrino E, Arcidiacono M, Francini A, Ercoli L. Arbuscular mycorrhizal fungi with contrasting life-history strategies differently affect health-promoting compounds in field-grown tomato by changing arbuscule occurrence and mycorrhizal assemblages in roots. Biology and Fertility of Soils. 2024;**60**:115-136. DOI: 10.1007/s00374-023-01778-6

[69] Trejo D, Sangabriel-Conde W, Gavito-Pardo ME, Banuelos J. Mycorrhizal inoculation and chemical fertilizer interactions in pineapple under field conditions. Agriculture. 2021;**11**(10):934. DOI: 10.3390/agriculture11100934

[70] Campos-Mota L, Baca-Castillo GA, Contreras DJ, Lúa AM, Hernández RA. Fertigation and mycorrhiza in red raspberry grown in tepetate. Agrociencia. 2004;**38**(1):75-83. Available from: https://www.redalyc.org/pdf/302/30238108.pdf [Accessed: May 21, 2022]

[71] Golubkina N, Amagova Z, Matsadze V, Zamana S, Tallarita A, Caruso G. Effects of arbuscular mycorrhizal fungi on yield, biochemical characteristics, and elemental composition of garlic and onion under selenium supply. Plants. 2020;**9**(1):84. DOI: 10.3390/plants9010084

[72] Yang F, Pan Y, Ali A, Zhang S, Li X, Qi X, et al. Agronomic biofortification of garlic through selenium and arbuscular mycorrhizal fungi application. Horticulturae. 2021;**7**(8):230. DOI: 10.3390/horticulturae7080230

[73] Pabitra A, John F. Meta-analysis: Impact of AMF colonization on allium species. Archives of Agriculture Research and Technology. 2024;**5**:1073. DOI: 10.54026/AART/1073

[74] Rozpądek P, Rąpała-Kozik M, Wężowicz K, Grandin A, Karlsson S, Ważny R, et al. Arbuscular mycorrhiza improves yield and nutritional properties of onion (*Allium cepa*). Plant Physiology and Biochemistry. 2016;**107**:264-272. DOI: 10.1016/j.plaphy.2016.06.006

[75] Díaz-Franco A, Alvarado-Carrillo M, Ortiz-Chairez F, Grageda-Cabrera O. Plant nutrition and fruit quality of pepper associated with arbuscular mycorrhiza in a greenhouse. Revista Mexicana de Ciencias Agrícolas. 2013;**4**(2):315-321. DOI: 10.29312/remexca.v4i2.1251

[76] Mendoza-Villarreal R, Robledo-Torres V, Pérez-Rodríguez MÁ, Guillén-Enríquez RR, Martínez-Cueto V, Paredes-Jácome JR. Impact of cover, ecotype and endomycorrhiza on morphology and quality of piquín chili. Revista Mexicana de Ciencias Agrícolas. 2021;**12**(2):193-204. DOI: 10.29312/remexca.v12i2.2847

[77] Vázquez-Hernández MV, Arévalo-Galarza L, Jaen-Contreras D, Escamilla-García JL, Mora-Aguilera A, Hernández-Castro E, et al. Effect of *glomus mosseae* and *Entrophospora colombiana* on plant growth, production, and fruit quality of 'Maradol' papaya (*Carica papaya* L.). Scientia Horticulturae. 2011;**128**(3):255-260. DOI: 10.1016/j.scienta.2011.01.031

[78] Benkebboura A, Zoubi B, Akachoud O, Gholam C, Qaddoury A. Field inoculation of arbuscular mycorrhizal fungi improves quality attributes and yield of melon (*Cucumis melo*). The Journal of Horticultural Science and Biotechnology. 2023;**99**:1-17. DOI: 10.1080/14620316.2023.2272150

[79] Miceli A, Vetrano F, Torta L, Esposito A, Moncada A. Effect of mycorrhizal inoculation on melon plants under deficit irrigation regimes. Agronomy. 2023;**13**(2):440. DOI: 10.3390/agronomy13020440

[80] Cosme M, Franken P, Mewis I, Baldermann S, Wurst S. Arbuscular mycorrhizal fungi affect glucosinolate and mineral element composition in leaves of *Moringa oleifera*. Mycorrhiza. 2014;**24**:565-570. DOI: 10.1007/s00572-014-0574-7

[81] Gashgari R, Selim S, Abdel-Mawgoud M, Warrad M, Habeeb TH, Saleh AEH. Arbuscular mycorrhizae induce a global metabolic change and improve the nutritional and health benefits of pennyroyal and parsley. Acta Physiologiae Plantarum. 2020;**42**:102. DOI: 10.1007/s11738-020-03091-3

[82] Avio L, Sbrana C, Giovannetti M, Frassinetti S. Arbuscular mycorrhizal fungi a effect total phenolics content and antioxidantactivity in leaves of oak leaf lettuce varieties. Scientia Horticulturae. 2017;**224**:265-271. DOI: 10.1016/j.scienta.2017.06.022

[83] Huang Z, Wang L, Meng S, Song X, Long R, Huang H, et al. The physiological responses of celery (*Apium graveolens* L.) and its ability to accumulate selenium when inoculated with *Funneliformis mosseae*. Scientia Horticulturae. 2024;**326**:112752. DOI: 10.1016/j.scienta.2023.112752

[84] Zeng L, Li J, Liu J, Wang M. Effects of arbuscular mycorrhizal (AM) fungi on citrus fruit quality under nature conditions. Southwest China Journal of Agricultural Sciences. 2014;**27**(5):2101-2105. DOI: 10.16213/j.cnki.scjas.2014.05.067

Chapter 3

Perspective Chapter: Symbiotic Interactions among Plant Pests and Pathogens – Nematodes, Bacteria, Viroids, Viruses, Insects, and Other Organisms

Refik Bozbuga, Ümit Haydar Erol, Bekir Bülent Arpacı, Pakize Gök Güler, Pinar Arıdıcı Kara, Hatice Nilufer Yıldız and Doğancan Kahya

Abstract

Several biotic and abiotic stress conditions adversely affect plant development. Nematodes, viruses, insects, viroids, bacteria, fungi, and numerous other pests are biotic stressors that significantly impact plant health and growth. These organisms may also establish symbiotic relationships with each other and their hosts, which can be either parasitic or mutualistic. This book chapter reviews some of these interactions, particularly focusing on the relationships among plant pests, other organisms, and their association with viruses and viroids under abiotic stress conditions. By detailing these symbiotic relationships between various hosts, this chapter aims to elucidate pest behavior and the complex dynamics involved. Understanding these interactions among plant pests, including nematodes, bacteria, viroids, viruses, insects, and other organisms, is crucial for developing effective pest management strategies and enhancing plant health. Future research should reveal molecular mechanisms of these interactions, explore pest resistance to climate change, and develop integrated management strategies for improved plant resistance.

Keywords: nematodes, insects, bacteria, viruses, Viroids, symbiotic interactions

1. Introduction

Symbiosis is defined as a relationship based on the mutual benefit of coexisting organisms, and various symbiotic relationships can be found between different species [1]. Furthermore, symbiosis is a close relationship that includes mutualistic, which is an interaction between two or more species where each species has a benefit, parasitic, in which one benefits from the other, and commensal associations in which at least two unrelated species exchange materials, energy, or information in a mutually beneficial manner [2, 3]. It is generally assumed that symbiotic organisms co-evolve and depend on each other to sustain their life cycles [4]. Microorganisms, extensively spread in the environment and colonizing nearly all ecological niches, must coinhabit certain habitats and interact, leading to the production of secondary metabolites that enable coexistence and survival, with bacteria being well-organized colonizers of plant tissue surfaces or interiors [5]. Due to limited resources, plant-associated bacteria and other microorganisms interact with each other, leading to complex interactions that enhance the survival, spread, and impact of plant pests, often to the detriment of the host plants [6]. In addition to the above, many living organisms engage interactions with one another. This book chapter reviews these symbiotic interactions, focusing on relationships involving insects, nematodes, bacteria, viroids, and viruses.

2. The symbiotic relationship between insects

A significant group of organisms where symbiotic systems are prevalent is insects that exhibit significant diversity and play crucial roles in ecosystem services, notably as pollinators and agents of pest control [7]. Several researches indicate that insects engage in various relationships with numerous macro- and microorganisms.

2.1 The symbiotic relationship between hemipteran insects and ants

One of the first symbiotic relationships encountered among symbiotic interactions is the mutualistic relationship between honeydew-producing insect species and ants. Among these insect species, whiteflies, scale insects, mealybugs, aphids, treehoppers, planthoppers, psyllids, and leafhoppers are notable, and their association with ants is significant in terms of both ecosystem dynamics and pest management that particularly, the sugary substance produced by these pest insects as they feed on plants constitutes a food source for ants [8, 9]. Ants obtain 62% of their nutrition from honeydew with 33% derived from prey insects and 5% from sources such as resin, fungi, dead insect remains, and seeds [8]. The relationship between ants and honeydew-producing insects, such as ants and scale insects worldwide, is believed to have started around 15–20 million years ago [10].

The honeydew-producing hemipteran insects, which feed by extracting sap from host plants and excreting the excess, produce a carbohydrate-rich substance known as honeydew or variably as "manna"; in warm and humid conditions, this substance forms a thin film on plants, facilitating the growth of saprophytic fungi, typically sooty mold, which creates a black layer that obstructs photosynthesis, inhibits plant growth, causes leaf and fruit drop, and diminishes the plant's esthetic and market value [11]. On a family level, a symbiotic relationship based on mutual benefit exists between honeydew-producing hemipteran species and ants that studies have identified certain ant species such as *Linepithema humile* (Mayr), *Pheidole megacephala* (Fabricius), and *Solenopsis* species living in association with hemipterans [9].

Whitefly nymphs secrete honeydew, which attracts ant species [12, 13]. For ant colonies, the proteins and lipids present in honeydew are important [14]. Ants reduce the risk of damage to whitefly nymph groups from both fungal infections and parasitoids, thereby increasing the survival rate of whiteflies [15].

Ants engage in various protective behaviors against predators and parasitoids, thereby maintaining a specific relationship with coccoids and other honeydew-producing hemipteran insects, by directly consuming the eggs, larvae, and adults of predators and indirectly impeding the feeding and oviposition of both parasitoids and predators [9]. The ant's response varies depending on the proximity and abundance of coccoids and the accessibility of other food sources [16]. When ants are scarce in cultivated crops, parasitoid and predator populations can control coccoid populations. However, the presence of ants reduces parasitism rates by disturbing parasitoids in the coccoid population [9]. A study conducted in Hawaii found that the presence of ants led to a faster increase in *C. viridis* populations and reduced mortality rates [17]. Additionally, ants have been observed to repel *C. montrouzieri* and *Azya orbigera* Mulsant (Coleoptera) larvae [16]. Some predators may escape ant detection due to their slow movement, hiding behavior, or pesticide application [16]. The disturbance caused by ants to parasitoids is correlated with the parasitoid's oviposition period; as the oviposition period increases, the likelihood of disturbance also increases [18]. Studies have shown that depending on the parasitoid species disturbed by ants, the parasitization rate of coccoids can decrease from 98.4 to 27.4% [9]. The red imported fire ant, *Solenopsis invicta*, native to South America, is an invasive species with significant ecological impacts in the United States, such as reduced biodiversity and disruption of native ant communities [19, 20]. In South China, *S. invicta* negatively affects agriculture, forestry, human health, and poultry production [21]. Although native ants also form mutualistic relationships with *P. solenopsis*, *S. invicta* is more effective in obtaining honeydew and thus plays a more dominant role [22].

The ghost ant *Tapinoma melanocephalum* and the invasive mealybug *Phenacoccus solenopsis* have a close mutualistic relationship that honeydew produced by *P. solenopsis* attracts *A. bambawalei*; honeydew consumption by ghost ants negatively affects parasitoid performance, and pygidial gland secretions and visual cues from ghost ants significantly inhibit parasitism [23].

2.2 The relationship between insects and mites

Insects engage extensively with mites, profoundly influencing insect populations, as documented in various ecological and evolutionary studies, with notable examples including quantitatively described mutualisms between mites and insects [24, 25]. Coevolutionary relationships exist between chaetodactylid mites and carpenter bees and the use of mites as biological control agents in agriculture through parasitism or predation [26]. Despite these interactions, mites typically exert minimal influence on insect population control under natural conditions due to their small size, which often preclude effective predation on living larval or adult insects; instead, insects play a crucial role in the dispersal (phoresy) and parasitism of mites [27, 28].

Mites, both scavengers and parasites, are among the most prevalent and diverse inhabitants of temporary bivouacs and nests [29]. Certain mite species exhibit specific morphological adaptations that facilitate their symbiotic roles with ants; for instance, some members of the Macrochelidae family are parasitic to *Eciton burchellii* and related species, attaching to membranous areas of the host's body and feeding on its blood [30]. Although mutualism has been proposed due to the morphological adaptation of the mite *Macrocheles rettenmeyeri* Krantz, which attaches specifically to the leg membrane of *Eciton dulcius* crassinode Borgmeier, and the ant's unusual tolerance potentially resulting from the mite's chemical mimicry [31], no clear benefit to the ant in this relationship has been observed [32].

2.3 The symbiotic relationship between insects and microorganisms

Symbiotic relationships can be found between insect and fungal species [33]. While symbiotic relationships with fungi are observed in many insect groups, this relationship highly evolved in ants (Hym.: Formicidae, Myrmicinae, Attini), termites (Isop.: Termitidae, Macrotermitinae), and Ambrosia beetles (Col.: Curculionidae, Scolytinae, and Platypodinae) [34]. Approximately 220 species of ants, primarily from the subfamily Myrmicinae, have been identified as engaging in symbiotic relationships with fungi [35]. In termites, around 330 species in the subfamily Macrotermitinae form symbiotic relationships with fungi of the genus Termitomyces [33].

The subfamily Scolytinae, which includes bark beetles that reproduce under the bark and ambrosia beetles that reproduce in wood, is divided into two main groups [36]. Ambrosia beetles, which predominantly establish galleries within the sapwood of dead or dying plants, introduce and subsist on symbiotic fungi; while most species are mycetophagous, relying exclusively on these fungi for nourishment at both larval and adult stages, xylomycetophagous species additionally digest portions of xylem tissue along with the fungi [37]. This symbiotic relationship ensures the continued survival of both ambrosia beetles and ambrosia fungi [38].

3. Nematode symbiosis with other organisms

Nematodes, thread-like and bilaterally symmetrical unsegmented animals often unseen due to their size and hidden habitats, inhabit diverse environments and, as parasites of plants, animals, and humans, significantly impact agriculture and economies, causing substantial annual losses in crops, forests, and other areas [39]. The nematode-bacteria symbiosis is significant and carries importance for both groups of organisms that nematodes, having evolved diverse bacterial associations ranging from mutualistic to pathogenic, serve as exemplary models for studying host-microbe interactions; recent genome-scale studies have elucidated communication mechanisms and enhanced understanding of genome evolution, mutualism, and innate immunity, offering broader insights applicable to other biological systems, including human diseases [40]. The symbiotic relationship between Steinernematidae and Heterorhabditidae nematodes with *Photorhabdus* and *Xenorhabdus* bacteria, which reside in nematode organs and produce antimicrobial metabolites, defines them as entomopathogenic nematodes; recent research highlights variations and additional bacterial contributors, enhancing specificity and advancing understanding of entomopathogen nematode-associated microbiota [41]. Obligate symbioses have shaped eukaryotic evolution, exemplified by the early-stage obligate symbiosis in *Howardula aoronymphium*, a nematode-parasitizing Drosophila, wherein a newly discovered maternally inherited intracellular bacterial symbiont, belonging to the Candidatus *Symbiopecto bacterium* group and exhibiting early genomic erosion, significantly aids nematode infection success, indicating a recent symbiotic relationship compared to other nematode species [42]. Heterorhabditis and Steinernema show a unique obligate symbiosis with the bacteria *Photorhabdus* and *Xenorhabdus*, crucial for insect host lethality and digestion, underpinning their role in biological pest control, and advancing research in entomopathogen nematode biodiversity, biogeography, and molecular genetics [43].

A defensive symbiont, *Enterococcus faecalis*, on the microbiota composition of the nematode host *Caenorhabditis elegans*, revealing that while *E. faecalis* maintained pathogen resistance and increased *Pseudomonas* abundance, it minimally disrupted overall microbiota diversity, demonstrating the nuanced role of evolving bacterial symbionts in shaping nematode-associated microbial communities, and highlighting their potential for pathogen defense without compromising microbial balance [44]. The symbiotic relationship between the nematode *Neoaplectana carpocapsae* and its bacterium *Xenorhabdus nematophilus* using *Hyalophora cecropia* pupae reveals that while bacterial resistance to insect immunity is significantly enhanced within the nematode host due to nematode-excreted immune inhibitors, the nematodes themselves remain unaffected by insect immunity, thus underscoring the survival advantages conferred by nematode-bacterial symbiosis [45]. Entomopathogenic nematodes, facilitated by mutualistic Gram-negative bacteria during their infective juvenile stage, exhibit remarkable infectivity across insect species, subverting insect immune responses and influencing nematode growth, development, and reproduction, highlighting a complex interplay of regulatory mechanisms and molecular components essential for understanding parasitic strategies relevant to human infectious diseases [46].

4. Bacteria and symbiont interactions

Bacteria form symbiotic associations within plant ecosystems [47]. Endophytes, which can produce bioactive secondary metabolites with antimicrobial properties, increase the production of different compounds as the number of coexisting species in a plant rises, providing protection against phytopathogens, with various plant tissues being simultaneously colonized by different species, such as fungal endophyte hyphae colonized by endobacteria [48]. Bacteria produce antibiotics and toxins as part of their symbiotic associations [49]. Thus, complex microbial interaction patterns occur within plants, where the colonization of endophytic fungal mycelia by bacteria can result in biofilm formation and the production of antimicrobial substances such as indoleacetic [50].

Bacteria may be involved in symbiotic interactions with nematodes, insects, and fungi.

4.1 Bacteria and nematode symbiotic interactions

The interaction between bacteria and nematodes, which may have been extensive in the evolutionary past, is a diverse and dynamic field with significant implications for agriculture, ecology, and biotechnology, particularly in the context of extant interactions between bacteria and plant-parasitic nematodes [51]. Plant-parasitic nematodes and Cladoceran crustaceans are infected by endospore-forming bacterial pathogens *Pasteuria* spp. [52]. *Bacillus thuringiensis* has reported to have nematicidal effects against a wide range of nematodes [53]. Root-knot nematodes (*Meloidogyne* spp.) carry endosymbiotic bacteria that enhance the nematodes' ability to infect plant roots and suppress plant defense mechanisms, benefiting from improved infection efficiency while providing the bacteria with a protected niche and access to plant-derived nutrients; these interactions are leveraged to develop genetic tools and assays for studying microbial pathogenicity, symbiosis, and immune responses [54]. Some plant-parasitic nematodes carry Burkholderia bacteria that produce enzymes to degrade plant cell walls, aiding the nematodes in penetrating plant tissues and accessing nutrients, while the bacteria benefit from the association by gaining access to plant tissues they might not independently reach [51]. Nematodes in the genera Steinernema and Heterorhabditis have mutualistic relationships with *Xenorhabdus* and *Photorhabdus* bacteria, respectively, with these bacteria aiding the nematodes in infecting and killing insect hosts, thereby providing a food source for both the nematodes and the bacteria [55].

4.2 Insects and bacteria symbiotic interactions

Entomopathogenic bacteria, which infect insects, have successfully been used along with their toxins in biological control programs as environmentally safer alternatives to synthetic pesticides [56]. Bacteria belong to species from different families such as Bacillaceae, Pseudomonadaceae, Enterobacteriaceae, Streptococcaceae, Micrococcaceae, Paenibacillaceae, Neisseriaceae, Rickettsiaceae, and Coxiellaceae are known as most common entomopathogenic organisms [57]. Entomopathogenic bacteria employ various mechanisms of action, predominantly by secreting toxic compounds into the hemocoel that suppress the insect's immune responses, thus enhancing their insecticidal potential [58]. Phytoplasmas are obligate symbiotic bacteria that infect plants and insects, typically requiring both hosts for dispersal, with transmission primarily facilitated by phloem feeders such as leafhoppers (Cicadellidae), planthoppers (Fulgoromorpha), and psyllids (Psyllidae) [59]. They can accumulate inside and outside insect cells [60]. Phytoplasmas must traverse the gut and salivary gland cells to reach the saliva for subsequent introduction into the phloem during insect feeding [61, 62]. Approximately 7 days after being introduced into a plant by an insect vector, phytoplasma symptoms may develop over a period ranging from 6 to 24 months, depending on the phytoplasma and plant species; infected plants, exhibiting symptoms such as yellowing and stunting, become more suitable for leafhopper feeding and reproduction, thereby facilitating the spread of phytoplasmas by these mobile vectors, while the bacterial wilt pathogen *Ralstonia solanacearum* can similarly be transmitted by beetles feeding on infected plants [63].

4.3 Fungi and bacteria symbiotic interactions

Bacteria and fungi share microhabitats by forming co-evolving communities whose interactions play a key role in ecosystem processes, vital in microbiology and ecology, impacting various environments and applications, including soil health [64]. Symbiotic interactions between bacteria and fungi are common in nature, spanning diverse fields such as ecology, agriculture, medicine, and biotechnology, and are critical for understanding ecosystem functions, improving plant health, developing new medical treatments, and creating innovative biotechnological applications; these interactions, which can be symbiotic (e.g., mycorrhizal fungi and nitrogen-fixing bacteria) or antagonistic (e.g., bacteria-producing antifungal compounds), play crucial roles in nutrient cycling, organic matter decomposition, and soil structure maintenance, particularly in the rhizosphere where they promote plant health by facilitating nutrient absorption and protecting against pathogens [65]. Mixed cultures of bacteria and fungi are used in fermentation processes to produce food, beverages, antibiotics, and enzymes. In some cases, bacteria and fungi can work together to degrade environmental pollutants, enhancing bioremediation efforts [66].

5. Viroids and abiotic stress beneficial effects

Viroids are the smallest pathogens that cause diseases in plants and have attracted significant interest in the field of plant pathology [67]. These pathogens, composed entirely of RNA molecules, lack the capacity to code for proteins, yet they induce significant morphological, physiological, biochemical, and molecular changes in their plant hosts [68]. Viroids previously were known to have had only devastating effects on plant health, and several recent studies have revealed that the interactions between the viroid pathogens and the host plant are more complicated [69]. Plant growth seed breeding also depended on viroids, viruses, bacteria, and fungi that the plant against viroid infections has recoverable genes that would make them resistant to abiotic hits, such as salinity, heat, and drought, among others [70].

5.1 Effects of viroids on plants

Viroids cause significant changes in plants, leading to deformation without a specific pattern and incomplete development of the plant's structures [71]. For instance, plant cells are much more orderly now with the reduction of tissues and longer roots that lead to a more protective plant against abiotic stress [72]. Viroid infections are no less than the prime mover of these changes in the leaves, roots, and stems of plants, which causes an increase in their water retention capacity, nutrient uptake, and general stress tolerance [73]. Leaf thickening often occurs due to diseases like viroids, which help the plant save water and increase drought resistance [74]. The overgrowth of roots may cause the root systems to unfold away from each other and the soil, and more healthy root system with a better ability for both water and nutrient uptake, and hence a stronger resistance to both salinity and drought stress [75]. The leaves that curl (**Figure 1**) and shrink save some leaf surface, a move that minimizes evaporation, while the plant gets better drought tolerance [76]. Increased stomatal density, a result of viroid infections, helps plants take advantage of a better control of transpiration, consequently boosting water use efficiency and strengthening the defense against the situation of drought, but the stem gets thicker for mechanical and heat stress resistance against high temperatures [77].

Viroid infections cause the changes in life processes of plants (**Figure 2**). They can actually change the function of plants in a way, which can allow them to adapt to water stress, without closing the stomata, elongate water usage, and alter the rate of photosynthesis [78]. For example, in maize (*Zea mays*) plant, viroid infection may cause the stomata to stay closed, helping the plant conserve water and use it more efficiently [79]. Viroid infection in barley (*Hordeum vulgare*) can enhance water use efficiency and provide the plants with drought tolerance [80]. But viroid infection in cotton (*Gossypium hirsutum*) can make the leaves close their stomata much quickly, besides this water loss; but, when the conditions get turned into drought, the plants can adapt better to the climate by inducing more stomatal closure [81].

Viroid infections similarly change the chemical makeup of the plants, leading to the activation of more of the antioxidant enzymes and, as a consequence, the plant's stronger resistance to oxidative stress [82]. The infection of plants with viroids is also a process that leads to the production of phenolic compounds in plants, which makes them less sensitive to the dramatic effects of environmental stress factors [83]. In tomato (*Solanum lycopersicum*), another example of a plant subjected to viroid infection, antioxidant enzyme activities will increase, and consequently, the plant will show better resistance to oxidative stress [84]. The synthesis of the production

Figure 1.
Symptoms caused by Potato spindle tuber viroid *(PSTVd) on the H5656 tomato variety (Photographed by Pakize Gök Güler).*

Figure 2.
Symptoms caused by Citrus exocortis viroid *(CEVd) on the H5656 tomato variety (Photographed by Pakize Gök Güler).*

of phenolic compounds of a pepper plant (*Capsicum annuum*) is viroid infection that boosts the plant's phenol level, which becomes the resistance factor to environmental stress [85]. Viroid infection raises the level of proline in the fertilizer that helps to maintain osmotic balance and to resist the drought stress in potato plants (*Solanum tuberosum*) [86].

At the gene level, viroid infections alter the expression patterns of plant genes [87]. The stress-related genes expression appears to be particularly the one of the genes that are mostly expressed, allowing plants to obtain the resistance against abiotic stress factors [88]. Viroid infection promotes the expression of stress-related genes; thereby, the tomato plant's resistance to drought stress is enhanced [89]. The infection of viroid in pepper (*Capsicum annuum*) changes the gene expression profiles in response to heat stress, and the tolerance is being adjusted [83]. Infected by viroid, plants regulate gene expression patterns that may lead to better stress tolerance [90].

5.2 Effects of viroids on plants under abiotic (salt, heat, drought) stress factors

As salt stress is the most common abiotic factor, it has a serious impact on the physiological and biochemical processes in plants by reducing water uptake in the roots and disrupting ion homeostasis, which leads to osmotic stress and causes cellular damage and decreases growth [91]. Plant defense mechanisms responding to moderate salt concentration sigmoid might include the involvement of viroids and plant adaptation to salinity [92]. One of the examples is the overexpression of proline, due to the viroid infection in tomato plants the function of which is regulating osmotic potential and water transport, and as a result, can alleviate the negative effects of salt stress [93]. Likewise, viroid infections in sunflower plants (*Helianthus annuus*) were presented to initiate the activity of the key antioxidant enzymes that are necessary to eliminate reactive oxygen species under salt stress and, therefore, give plants more resistance to saline conditions [94].

Heat stress is an important abiotic factor and is the main cause of plant growth suppression and developmental disorders [95]. Exposing plants to high temperatures can make their proteins to change their shape, modify their cell membranes, and reduce the efficiency of their cellular photosynthesis, leading to the overall weakening and deterioration of the plants [96]. Nevertheless, modern research has revealed the fact that viroid infections can, within certain contexts, show a new side of plants that strengthens their heat-resistant abilities by activating some other defense mechanisms that pave the way for them to adapt to the elevated temperatures [97]. The infecting viroid is even observed to be able to turn on the heat shock proteins (HSPs) synthesis that is the protein in the form of molecular chaperones able to function as a heat shock preventing protein denaturation and as a refolded damaged protein, so that the cell could be homeostatic again under the heat stress [98]. For example, viroid infection produces a plethora of genes that are responsible for the increased expression of heat-shock genes in tobacco plants, allowing the plant to withstand prolonged high temperature [99]. Additionally, viroid infections have been known to increase the antioxidant defense system in a plant that is necessary to keep oxidative damage in check due to heat stress [100]. Cucumber plants packed in the genus *Cucumis sativus* get an infection with the viroid, and this causes the plants to give off the superoxide dismutase (SOD) and catalase (CAT) that are the enzymes cleaning the oxidative stress that has been triggering the reactive oxygen species (ROS) in the plants [101]. Another example is in the grapevine in *Vitis vinifera* infected with a viroid has increased the synthesis of compatible solutes such as trehalose and glycine betaine, which enable the proteins and membranes of the plant to resist heat-induced denaturation [102].

Drought stress is one of the main abiotic stress factors, which keeps the plant away from the water that the plant needs so much, obstructing the growth process and thereby damaging it [103]. This stress can impair chemical processes essential for plant survival and growth, thus reducing plant vigor, lowering yield, and ultimately causing plant death [104]. Viroid infections cause changes in physical, physiological, biochemical, and molecular properties that support plant adaptation to drought stress [105].

The perfect example of the viroid infection's impact on the tomato plants is the study conducted by Mihaljević et al. [106], in which uninfected tomatoes showed high levels of abscisic acid (ABA), one of the examples helping the plant regulate the opening and closing of stomata and therefore save water; ABA, which regulates stomatal opening and closing, is high, and this is the only mechanism that will enable the plant to withstand dry periods [107]. A viroid mechanism induces the biosynthesis of osmoprotectants such as proline, trehalose, and soluble sugars, which facilitate the maintenance of turgor and maintenance of structure [107]. For example, research has been providing evidence that viroid-infected cucumber plants (*Cucumis sativus*) are able to store more proline to act as an osmoprotectant and consequently withstand dehydration by stabilizing proteins and membranes [108].

6. The influence of environmental stresses, viruses on plant physiology, and the plant-virus interactions

6.1 Effects of salt stress and viral infections on plant physiology

Plants are affected by biotic and abiotic stresses; abiotic stresses such as temperature fluctuations, drought, salinity, and other environmental factors can have an adverse effect on the physiological and biochemical processes of plants [109]. In particular, the decrease of sodium chloride is one of the most crucial abiotic stresses that causes excess of salt stress to plants and limits their preference for water and nutrients, thereby very much lowering their growth and developmental processes [110].

Upon entering plant cells, viral infections get adapted to the cell environment where they alter cellular functions resulting in various symptoms in plants such as yellowing of the leaves (**Figure 3**), stunted growth, necrosis, and deformation [111]. Viruses halt normal physiological processes through the brainwash of plant cell metabolism, the expression of gene, and the protein homeostasis [112]. The viral infections of green plants can also raise its vulnerability to salt stress, while the virus has all the time been combating the plant's defense factors, mainly [113]. Most virus-induced generation of reactive oxygen species (ROS) is associated with the activation of various signaling molecules and transcription factors [114]. Additionally, viruses can alter gene expression in plant cells, reducing the plant's ability to produce defense proteins, making plants more susceptible to both viral infections and abiotic stresses [115].

Salt stress is a common problem in agriculture, leading to adverse effects such as osmotic stress, ion toxicity, and nutrient imbalances in plants and that plants exposed to salt stress exhibit various physiological disorders, including stunted growth, chlorosis, and even death [116]. Salt stress prevents water intake and disrupts the ionic balance in cells, leading to metabolic dysfunction [117]. High salt concentrations especially affect root development and water uptake, reducing the overall growth rate of plants [117]. The effects of salt stress on plants can vary depending on the genetic makeup of the plant and environmental conditions [118]. Some plant species are more tolerant to salt stress and can maintain normal growth and developmental processes even under high salt concentrations [119].

6.2 Interactions between viral infections and salt stress

Interactions between viral infections and salt stress can sometimes lead to beneficial outcomes for plants by triggering stress responses that enhance plant tolerance to

Figure 3.
Symptom caused by Beet western yellows virus *on a pepper plant in the field (Photographed by Bekir Bülent Arpacı).*

salt stress [120]. For example, viruses can induce the production of osmoprotectants such as proline and glycine betaine, which help plants maintain cellular osmotic balance under salt stress [121]. Certain viral infections can also enhance the antioxidant defense systems of plants, reducing the damage caused by ROS under salt stress [122]. Moreover, some viruses can improve water uptake in salt-stressed plants by modulating aquaporin expression, helping maintain cellular water balance [123]. Viral infections can enhance the expression of stress-responsive genes that contribute to improved salt tolerance [124]. By manipulating cellular signaling pathways, viruses can activate pathways that enhance the plant's ability to cope with salt stress, promoting growth and development even under adverse conditions [125]. In tomato plants, the co-occurrence of salt stress and *Tomato Yellow Leaf Curl Virus* (TYLCV) infection has been shown to induce the production of stress-related proteins that improve salt tolerance [126]. In barley, infection with *Barley Yellow Dwarf Virus* (BYDV) under salt stress conditions was found to increase the antioxidant capacity of plants and reduce oxidative damage [127]. Rice plants exposed to both salt stress and *Rice Tungro Virus* (RTV) infection show improved water retention and reduced osmotic stress due to virus-induced expression of aquaporins [128]. In wheat, infection with *Wheat Streak Mosaic Virus* (WSMV) can lead to the upregulation of genes involved in salt stress tolerance, helping plants to better withstand high-salinity conditions [129]. Tobacco plants infected with *Tobacco Mosaic Virus* (TMV) under salt stress conditions show enhanced antioxidant activity, reducing ROS accumulation and mitigating cellular damage [130]. In cucumber, infection with *Zucchini Yellow Mosaic Virus* (ZYMV) has been reported to enhance salt tolerance by modulating the expression of salt stress-responsive genes [131]. In grapevine, infection with *Grapevine Fanleaf Virus* (GFLV) under salt stress conditions can improve ion homeostasis and reduce sodium toxicity [132]. In soybean, infection with *Soybean Mosaic Virus* (SMV) was found to increase plant tolerance to salt stress by inducing the expression of stress-related proteins [133]. *Maize Dwarf Mosaic Virus* (MDMV)-infected maize plants, under salt stress conditions, can improve growth and yield by modulating hormonal pathways involved in stress responses [134].

6.3 The impact of drought stress on plant physiology and effects of viral infections on stress and plants

Drought stress significantly affects plant growth, development, and yield by influencing water uptake and retention capacity, leading to a decrease in plant water potential, slowing of cell division, and a reduction in photosynthesis rates [135]. Under drought conditions, plants activate various adaptation mechanisms to reduce water loss and enhance water uptake. At the cellular level, the production of osmolytes increases, maintaining intracellular water balance, and antioxidant defense mechanisms are activated [136, 137]. Physiological and morphological changes include stomatal closure and increased root development [138]. At the genetic and epigenetic levels, the expression of stress response genes changes, and epigenetic modifications such as DNA methylation occur [139]. Metabolic adjustments involve the reorganization of photosynthesis and carbon metabolism to enhance water use efficiency [140].

Viral infections typically have pathogenic effects on plants by affecting their physiological and biochemical processes [141]. Viruses can trigger plant defense mechanisms and cause disruptions in cellular functions. Plants develop various defense mechanisms against viral infections, including hypersensitive responses, RNA interference, and activation of defense genes [142]. After entering the cell, viruses manipulate plant metabolism to enhance their replication and spread, and this process disrupts normal cellular activities and generally reduces growth and development [143].

Defense mechanisms triggered by viruses in plant cells can lead to changes in the levels of plant hormones such as salicylic acid, jasmonic acid, and ethylene, which coordinate the plant's overall defense response, providing protection against both biotic and abiotic stresses [144]. The RNA interference (RNAi) mechanism helps plants recognize and target viral genetic material for degradation [145].

6.4 Symbiotic effects of viral infections and drought stress

Recent research has revealed that viral infections can have positive effects on plants under drought stress. For example, barley (*Hordeum vulgare*) plants infected with *Barley Yellow Dwarf Virus* (BYDV) performed better under drought conditions than uninfected plants due to the virus enhancing water use efficiency and regulating the expression of stress response genes [146]. Similarly, tomato plants infected with *Tomato Yellow Leaf Curl Virus* (TYLCV) exhibited lower oxidative stress levels and higher antioxidant enzyme activities under drought stress [147]. *Cucumber mosaic virus* (CMV)-infected tobacco (*Nicotiana tabacum*) plants are found to use less water and have increased antioxidant activity during drought stress [148]. *Bean Common*

Mosaic Virus (BCMV)-infected bean (*Phaseolus vulgaris*) plants also grow better under drought conditions [149]. Potato (*Solanum tuberosum*) plants, on the other hand, infected with Potato Virus Y (PVY) acquire increased drought tolerance and better water use efficiency [150].

Furthermore, zucchini (*Cucurbita pepo*) plants infected with *Zucchini Yellow Mosaic Virus* (ZYMV) showed increased drought resistance and lower oxidative stress levels [151]. Rice (*Oryza sativa*) plants infected with *Rice Tungro Bacilliform Virus* (RTBV) also reported increased resistance to drought stress and improved water use efficiency [152].

6.5 Viral enhancements of plant heat tolerance

High temperatures can disrupt the structure of plant cell membranes, denature proteins, and interfere with photosynthetic processes by impairing chloroplast and mitochondrial function, negatively affecting energy production and photosynthesis [153]. Moreover, elevated temperatures can induce oxidative stress, which in turn triggers the excessive production of ROS [154]. The overabundance of ROS can initiate lipid peroxidation in cell membranes, inflict damage on DNA, and cause protein oxidation, severely impairing cellular functions and leading to a marked decline in plant growth and productivity [155].

Research shows that viruses can enhance drought tolerance in plants by altering hormone signaling, antioxidant defenses, and gene expression [156]. Some viruses stimulate the production of heat shock proteins (HSPs) in plants, helping protect cells from heat stress [157]. Thus, it increases the survival and productivity of plants by making them more resistant to abiotic stresses [157].

Pepper (*Capsicum annuum*) plants infected with *Pepper Golden Mosaic Virus* (PepGMV) suffer less damage under high heat stress than uninfected plants, as the virus activates plant defense mechanisms, preserves cell structure, and regulates energy metabolism, making the plants more resilient to the negative effects of heat stress [158]. Similarly, *Turnip Mosaic Virus* (TuMV) infection is associated with increased tolerance to heat stress in Brassica plants by stimulating the production of heat shock proteins (HSPs) and other defense mechanisms [159].

TYLCV infection strengthens antioxidant systems and activates cellular repair mechanisms, making tomato plants more resistant to high temperatures that similar viral infections in other plants such as almond (*Prunus dulcis*), maize (*Zea mays*), wheat (*Triticum aestivum*), barley (*Hordeum vulgare*), tobacco (*Nicotiana tabacum*), sugar beet (*Beta vulgaris*), cotton (*Gossypium hirsutum*), and potato (*Solanum tuberosum*) have been reported to trigger specific resistance mechanisms under heat stress [160–162].

In sugar beet and cotton plants, viral infections optimize cellular metabolism and reduce ROS levels, increasing plant resistance to heat stress [163]. In potato plants, viral infection stimulates the production of heat shock proteins and antioxidant enzymes, facilitating adaptation to high temperatures [164]. These mechanisms increase plant survival and productivity under heat stress, with positive implications for agricultural productivity [165].

7. Conclusions

Symbiotic relationships occur between many organisms, including insects, microorganisms, nematodes, bacteria, and fungi, and are important for ecological balance. Understanding of symbiotic interactions is of great importance in plant pest control

Symbiotic relationship type	Organisms involved	Interaction type	Ecological impact	Agricultural significance
Nematode-bacteria	Nematodes (*Steinernema*, *Heterorhabditis*) and bacteria (*Xenorhabdus*, *Photorhabdus*)	Mutualistic parasitism	Reduces insect populations, controlling pest species in ecosystems	Used in biological pest control to reduce insect pests in agriculture
Viral infections	Plants (e.g., tomato, barley, cucumber) and viruses (e.g., TYLCV, BYDV)	Pathogenic symbiosis	Alters plant physiology to increase drought and heat resilience	Virus-infected plants can exhibit improved stress tolerance in crop systems
Insect-fungi	Bark beetles and ambrosia fungi	Mutualistic (Mycophagy)	Facilitates wood decomposition and nutrient recycling in forests	Detrimental in agriculture due to tree damage caused by beetle-fungi associations
Insect-hemiptera (ants-aphids)	Ants and hemipteran insects (aphids, whiteflies)	Mutualistic	Increases survival of hemipterans, disrupting natural predators	Ant-aphid interactions can lead to outbreaks of aphid populations in crops
Viroid-plant	Various plants (e.g., potato, tomato) and viroids	Pathogenic	Can induce beneficial morphological changes under abiotic stress	Viroids may reduce growth but increase tolerance to drought and salinity in some crops
Bacteria-fungi	Endophytic bacteria and fungi	Mutualistic	Enhances nutrient exchange and plant protection against pathogens	Endophytes improve plant resilience, contributing to increased yield and protection from diseases

*Table was prepared by Ümit Haydar Erol.

Table 1.
*Comparative analysis of symbiotic relationships and their ecological/agricultural impact.**

and plant health. In addition, the parasitic and symbiotic relationships of bacteria with other organisms and the impact of these relationships in terms of nutrient cycle and ecological balance are important.

Plants can be exposed to abiotic stress factors as well as biotic stress factors, and in many abiotic stress situations such as salinity, drought, and high temperature. Viruses and viroids can play a role in changing the resistance of these plants against abiotic stresses.

This book chapter reviews symbiotic relationships and abiotic stress factors between different organisms such as nematodes, bacteria, insects, viruses, and viroids, and is believed to be valuable in providing a different perspective on ecosystem dynamism, structure, and relationship with ecosystems (**Table 1**). It is thought that it will shed light on future studies on controlling diseases and pests in plants and add a different perspective.

Author contribution

The contributions to this work are as follows: Nematode symbiosis with other organisms (by R. Bozbuga); The impact of abiotic stresses, viral infections on plant physiology, and their interactions (by B.B. Arpacı and Ü.H. Erol); Viroids and the beneficial effects of abiotic stress (by P.G. Guler); The symbiotic relationships between insects (by D. Kahya and P.A. Kara); and Bacterial and symbiont interactions (by H.N. Yildiz) have been authored in this book chapter.

Conflict of interest

The authors declare no conflict of interest.

Author details

Refik Bozbuga[1*], Ümit Haydar Erol[2], Bekir Bülent Arpacı[2], Pakize Gök Güler[3], Pinar Arıdıcı Kara[3], Hatice Nilufer Yıldız[3] and Doğancan Kahya[1]

1 Department of Plant Protection, Faculty of Agriculture, Eskisehir Osmangazi University, Eskisehir, Turkiye

2 Department of Horticulture, Faculty of Agriculture, Kilis 7 Aralık University, Kilis, Turkiye

3 Biological Control Research Institute, Adana, Turkiye

*Address all correspondence to: refikbozbuga@gmail.com

References

[1] Douglas AE. The Symbiotic Habit. Princeton (NJ): Princeton University Press; 2010. 202 p. ISBN 978-0-691-11341-8.

[2] Saffo MB. Coming to Terms with a Field: Words and Concepts in Symbiosis. Philadelphia/Rehovot: Balaban; 1992. pp. 17-31

[3] Chertow MR. Industrial symbiosis: Literature and taxonomy. Annual Review of Energy and the Environment. 2000;**25**:313-337. DOI: 10.1146/annurev.energy.25.1.313

[4] Beaver RA. Insect-fungus relationships in the bark and ambrosia beetles. In: Wilding N, Collins NM, Hammond PM, Webber JF, editors. Insect-Fungus Interactions. 14th Symposium of the Royal Entomological Society of London. London: Academic Press; 1989. 121 p. DOI: 10.1016/B978-0-12-751800-8.50011-2

[5] Bashir I, War AF, Rafiq I, Reshi ZA, Rashid I, Shouche YS. Phyllosphere microbiome: Diversity and functions. Microbiological Research. 2022;**254**:126888. DOI: 10.1016/j.micres.2022.126888

[6] Hassani MA, Durán P, Hacquard S. Microbial interactions within the plant holobiont. Microbiome. 2018;**6**(1):1-17. DOI: 10.1186/s40168-018-0445-0

[7] Okabe K. Ecological characteristics of insects that affect symbiotic relationships with mites. Entomological Science. 2013;**16**(4):363-378. DOI: 10.1111/ens.12050

[8] Way MJ, Khoo KC. Role of ants in pest managements. Annual Review Entomology. 1992;**37**:479-503. DOI: 10.1146/annurev.en.37.010192.002403

[9] Turguter S, Ülgentürk S. Karıncalar hymenoptera: Formicidae ve coccoidea hemiptera: Sternoryncha türlerinin ilişkileri. Journal of Agricultural Sciences. 2007;**13**(03):312-320. DOI: 10.1501/Tarimbil_0000000538

[10] Johnson CD, Agosti JH, Delabie K, Dumpert DJ, Williams M, Tschirnhaus M. von, Maschwitz U. Acropyga and Azteca ants with scale insects: 20 million years of intimate symbiosis. American Museum Noviates. 2001;**3335**:1-18. DOI: 10.1206/0003-0082(2001)335<0001:AAAAHF>2.0.CO;2

[11] Kosztarab M, Kozár F. Scale Insects of Central Europe. Vol. 41. Budapest: Akademiai Kiado; 1988. 442 p. DOI: 10.1007/978-94-009-4045-1

[12] DeBach P, Rose M. Biological control of woolly whitefly. California Agriculture. 1976;**30**:4-7

[13] Belay DK, Zewdu A, Foster JA. Ecology and management of the woolly whitefly (Hemiptera: Aleyrodidae), a new invasive citrus pest in Ethiopia. Journal of Economic Entomology. 2011;**104**:1329-1338. DOI: 10.1603/EC11017

[14] Delabie JHC, Mariano CSF, Serrão JE, Pompolo SG. The karyotype of the aberrant obligate ant-attended mealybug, *Neochavesia caldasiae* (Coccoidea: Pseudococcidae: Rhizoecinae). Caldasia. 2004;**26**:185-190. Available from: https://repositorio.unal.edu.co/handle/unal/73064

[15] Queiroz JM, Oliveira PS. Tending ants protect honeydew-producing

whiteflies (Homoptera: Aleyrodidae). Environmental Entomology. 2001;**30**(2):295-297. DOI: 10.1603/0046-225X-30.2.295

[16] Way MJ. Mutualism between ants and honeydew-producing Homoptera. Annual Review of Entomology. 1963;**8**:307-344. DOI: 10.1146/annurev.en.08.010163.001515

[17] Bach CE. Direct and indirect interactions between ants (*Pheidole megacephala*), scales (*Coccus viridis*), and plants (*Pluchea indica*). Ecology. 1991;**87**:233-239. DOI: 10.1007/BF00325261

[18] Bartlett BR. The influence of ants upon parasites, predators, and scale insects. Annals of the Entomological Society of America. 1961;**54**:543-551. DOI: 10.1093/aesa/54.4.543

[19] Porter SD, Savignano DA. Invasion of polygyne fire ants decimates native ants and disrupts arthropod community. Ecology. 1990;**71**:2095-2106. DOI: 10.2307/1938623

[20] Kaplan I, Eubanks MD. Aphids alter the community-wide impact of fire ants. Ecology. 2005;**86**:1640-1649. DOI: 10.1890/04-0016

[21] Zeng L, Lu YY, He XF, Zhang WQ, Liang GW. Identification of red imported fire ant *Solenopsis invicta* to invade mainland China and infestation in Wuchuan, Guangdong. Chinese Bulletin of Entomology. 2005;**42**:144-148

[22] Zhou A, Lu Y, Zeng L, Xu Y, Liang G. Does mutualism drive the invasion of two alien species? The case of *Solenopsis invicta* and *Phenacoccus solenopsis*. PLoS One. 2012;**7**(7):e41856. DOI: 10.1371/journal.pone.0041856

[23] Xu C, Li Q, Qu X, Chen J, Zhou A. Ant–hemipteran association decreases parasitism of *Phenacoccus solenopsis* by endoparasitoid *Aenasius bambawalei*. Ecological Entomology. 2020;**45**(2):290-299. DOI: 10.1111/een.12797

[24] Wilson DS, Knollenberg WG. Adaptive indirect effects: The fitness of burying beetles with and without their phoretic mites. Evolutionary Ecology. 1987;**1**:139-159. DOI: 10.1007/BF02067397

[25] Okabe K, Makino S. Parasitic mites as part-time bodyguards of a host wasp. Proceedings of the Royal Society B. 2008;**275**:2293-2297. DOI: 10.1098/rspb.2008.0586

[26] Gerson U, Smiley RL. Acarine Biocontrol Agents. New York: Chapman and Hall; 1990. 174 p. ISBN-13: 978-0412360602

[27] Athias-Binche F. Evolutionary ecology of dispersal in mites. In: Dusbábek F, Bukva V, editors. Modern Acarology. Vol. 1. The Hague: SPB Academic Publishing; 1991. pp. 27-41

[28] Houck MA, O'Connor BM. Ecological and evolutionary significance of phoresy in the Astigmata. Annual Review of Entomology. 1991;**36**:611-636. DOI: 10.1146/annurev.en.36.010191.003143

[29] Rettenmeyer CW, Rettenmeyer ME, Joseph J, Berghoff SM. The largest animal association centered on one species: The army ant *Eciton burchellii* and its more than 300 associates. Insectes Sociaux. 2011;**58**:281-292. DOI: 10.1007/s00040-010-0128-8

[30] Schmid-Hempel P. Parasites in Social Insects. Princeton (NJ): Princeton University Press; 1998. 392 p. ISBN: 9780691059242

[31] Eickwort GC. Associations of mites with social insects. Annual

Review of Entomology. 1990;**35**:469-488. DOI: 10.1146/annurev.en.35.010190.002345

[32] Krantz GW, Walter DE. A Manual of Acarology. 3rd ed. Lubbock (TX): Texas Tech University Press; 2009. 816 p. OCLC Number: 1422025157

[33] Kushiyev R, Aker O, Tuncer C. Ambrosia Böcekleri (Coleoptera: Curculionidae: Scolytinae ve Platypodinae) ile Ambrosya Fungusları Arasındaki Simbiyotik İlişkiler. Harran Tarım ve Gıda Bilimleri Dergisi. 2017;**21**(2):239-246. DOI: 10.29050/harranziraat.321179

[34] Mueller UG, Gerardo NM, Aanen DK, Six DL, Schultz TR. The evolution of agriculture in insects. Annual Review of Ecology, Evolution, and Systematics. 2005;**36**:63-95. DOI: 10.1146/annurev.ecolsys.36.102003.152626

[35] Schultz TR, Meier R. A phylogenetic analysis of the fungus-growing ants (Hymenoptera: Formicidae: Attini) based on morphological characters of the larvae. Systematic Entomology. 1995;**20**:337-370. DOI: 10.1111/j.1365-3113.1995.tb00100.x

[36] Selmi E. Türkiye Kabuk Böcekleri ve Savaşı. İstanbul: İstanbul Üniversitesi Yayın No: 4042, Emek Matbaası; 1998. 196 p

[37] Roeper RA. Patterns of mycetophagy in Michigan ambrosia beetles. Michigan Academician. 1995;**26**:153-161

[38] Harrington TC. Ecology and evolution of mycophagous bark beetles and their fungal partners. In: Vega FE, Blackwell M, editors. Ecological and Evolutionary Advances in Insect-Fungal Associations. New York: Oxford University Press; 2005. pp. 257-291. DOI: 10.1093/oso/9780195166521.003.0011

[39] Li Q, Liang W, Zhang X, Mahamood M, editors. Soil Nematodes of Grasslands in Northern China. 1st ed. London: Academic Press; 2017. 45-228 p. DOI: 10.1016/B978-0-12-813274-6.00003-X

[40] Murfin KE, Dillman AR, Foster JM, Bulgheresi S, Slatko BE, Sternberg PW, et al. Nematode-bacterium symbioses—Cooperation and conflict revealed in the "omics" age. The Biological Bulletin. 2012;**223**(1):85-102. DOI: 10.1086/BBLv223n1p85

[41] Tarasco E, Fanelli E, Salvemini C, El-Khoury Y, Troccoli A, Vovlas A, et al. Entomopathogenic nematodes and their symbiotic bacteria: From genes to field uses. Frontiers in Insect Science. 2023;**3**:1195254. DOI: 10.3389/finsc.2023.1195254

[42] Martinson VG, Gawryluk RMR, Gowen BE, et al. Multiple origins of obligate nematode and insect symbionts by a clade of bacteria closely related to plant pathogens. Proceedings of the National Academy of Sciences of the United States of America. 2020 Dec;**117**(50):31979-31986. DOI: 10.1073/pnas.2000860117

[43] Burnell A, Stock SP. *Heterorhabditis, Steinernema* and their bacterial symbionts—Lethal pathogens of insects. Nematology. 2000;**2**(1):31-42. DOI: 10.1163/156854100508872

[44] Dahan D, Preston GM, Sealey J, et al. Impacts of a novel defensive symbiosis on the nematode host microbiome. BMC Microbiology. 2020;**20**(1):159. DOI: 10.1186/s12866-020-01845-0

[45] Gotz P, Boman A, Boman HG. Interactions between insect immunity

and an insect-pathogenic nematode with symbiotic bacteria. Proceedings of the Royal Society of London - Series B: Biological Sciences. 1981;**212**(1188):333-350. DOI: 10.1098/rspb.1981.0043

[46] Eleftherianos I, Heryanto C. Molecular regulators of entomopathogenic nematode-bacterial symbiosis. Results and Problems in Cell Differentiation. 2020;**69**:453-468. DOI: 10.1007/978-3-030-51849-3_17

[47] Porras-Alfaro A, Bayman P. Hidden fungi, emergent properties: Endophytes and microbiomes. Annual Review of Phytopathology. 2011;**49**:291-315. DOI: 10.1146/annurev-phyto-080508-081831

[48] Hoffman MT, Arnold AE. Diverse bacteria inhabit living hyphae of phylogenetically diverse fungal endophytes. Applied and Environmental Microbiology. 2010;**76**:4063-4075. DOI: 10.1128/AEM.02928-09

[49] Boemare N. Interactions between the partners of the entomopathogenic bacterium nematode complexes, Steinernema-*Xenorhabdus* and Heterorhabditis-*Photorhabdus*. Nematology. 2002;**4**(5):601-603. DOI: 10.1163/15685410260438863

[50] Bandara WM, Seneviratne G, Kulasooriya SA. Interactions among endophytic bacteria and fungi: Effects and potentials. Journal of Biosciences. 2006;**31**(5):645-650. DOI: 10.1007/BF02708417

[51] Bird DM, Opperman CH, Davies KG. Interactions between bacteria and plant-parasitic nematodes: Now and then. International Journal for Parasitology. 2003;**33**(11):1269-1276. DOI: 10.1016/s0020-7519(03)00160-7

[52] Preston JF, Dickson DW, Maruniak JE, et al. Pasteuria spp.: Systematics and phylogeny of these bacterial parasites of phytopathogenic nematodes. Journal of Nematology. 2003;**35**(2):198-207

[53] Trivedi PC, Malhotra A. Bacteria in the management of plant-parasitic nematodes. In: Maheshwari D, editor. Bacteria in Agrobiology: Disease Management. Berlin, Heidelberg: Springer; 2013. DOI: 10.1007/978-3-642-33639-3_13

[54] Kennedy MW, Harnett W. Parasitic nematodes molecular biology, biochemistry and immunology. In: Kennedy MW, Harnett W, editors. Molecular Biology, Biochemistry and Immunology. Parasitic Nematodes: CABI Publishing; 2013. DOI: 10.1079/9781845937591.0000

[55] Ferreira T, Malan AP. *Xenorhabdus* and *Photorhabdus*, bacterial symbionts of the entomopathogenic nematodes Steinernema and Heterorhabditis and their in vitro liquid mass culture: A review. African Entomology. 2014;**22**(1):1-14. DOI: 10.4001/003.022

[56] Glare TR, Jurat-Fuentes J-L, O'Callaghan M. Basic and applied research: Entomopathogenic bacteria. In: Lacey LA, editor. Microbial Control of Insect and Mite Pests. London, UK: Academic Press; 2017. pp. 47-67. DOI: 10.1016/B978-0-12-803527-6.00004-4

[57] Gouli VV, Marcelino JAP, Gouli SY. Microbial pesticides: Biological resources, production and application. In: The Basic Biological Resources for the Production of Microbial Pesticides. London, UK: Academic Press; 2021. pp. 1-134. DOI: 10.1016/C2020-0-00269-0

[58] Hinchliffe SJ, Hares MC, Dowling AJ, ffrench-Constant RH. Insecticidal

toxins from the *Photorhabdus* and *Xenorhabdus* bacteria. The Open Toxinology Journal. 2010;**3**(1):83-100. DOI: 10.2174/1875414701003010083

[59] Weintraub PG, Trivellone V, Krüger K. The biology and ecology of leafhopper transmission of phytoplasmas. In: Bertaccini A, Weintraub P, Rao G, Mori N, editors. Phytoplasmas: Plant Pathogenic Bacteria - II. Singapore: Springer; 2019. DOI: 10.1007/978-981-13-2832-9_2

[60] Ammar E-D, Hogenhout SA. Mollicutes associated with arthropods and plants. In: Kostas B, Miller T, editors. Insect Symbiosis. 2nd ed. Boca Raton: CRC Press, Taylor and Francis Group; 2006. pp. 97-118. DOI: 10.1201/9781420005936

[61] Lefol C, Lherminier J, Boudon-Padieu E, et al. Propagation of flavescence doree MLO (mycoplasma-like organism) in the leafhopper vector *Euscelidius variegatus*. Journal of Invertebrate Pathology. 1994;**63**:285-293. DOI: 10.1006/jipa.1994.1053

[62] Lherminier J, Prensier G, Boudon-Padieu E, Caudwell A. Immunolabeling of grapevine flavescence doree MLO in salivaryglands of *Euscelidius variegatus*: A light and electron microscopy study. Journal of Histochemistry and Cytochemistry. 1990:**38**:79-85. DOI: 10.1177/38.1.2294149

[63] Remenant B, de Cambiaire JC, Cellier G, Jacobs JM, Mangenot S, Barbe V, et al. *Ralstonia syzygii*, the blood disease bacterium and some Asian R. *solanacearum* strains form a single genomic species despite divergent lifestyles. PLoS One. 2011:**6**(9):e24356. DOI: 10.1371/journal.pone.0024356

[64] Bastías DA, Johnson LJ, Card SD. Symbiotic bacteria of plant-associated fungi: Friends or foes? Current Opinion in Plant Biology. 2020;**56**:1-8. DOI: 10.1016/j.pbi.2019.10.010

[65] Tahat MM, Kamaruzaman S, R, Othman. Mycorrhizal fungi as a biocontrol agent. Plant Pathology Journal. 2010;**9**:198-207. DOI: 10.3923/ppj.2010.198.207

[66] Harrison DEF. Mixed cultures in industrial fermentation processes. Advances in Applied Microbiology. 1978;**24**:129-164. DOI: 10.1016/S0065-2164(08)70638-5

[67] Flores R, Di Serio F, Navarro B, Duran-Vila N, Owens R. Viroids and viroid diseases of plants. In: Hurst CJ, editor. Studies in Viral Ecology: Microbial and Botanical Host Systems. Vol. 1. Hoboken: Wiley; 2011. pp. 307-341. DOI: 10.1002/9781118025666.ch12

[68] Pfeilmeier S, Caly DL, Malone JG. Bacterial pathogenesis of plants: Future challenges from a microbial perspective: Challenges in bacterial molecular plant pathology. Molecular Plant Pathology. 2016;**17**(8):1298-1313. DOI: 10.1111/mpp.12427

[69] Hussain B. Modernization in plant breeding approaches for improving biotic stress resistance in crop plants. Turkish Journal of Agriculture and Forestry. 2015;**39**(4):515-530. DOI: 10.3906/tar-1406-176

[70] Nejat N, Mantri N. Plant immune system: Crosstalk between responses to biotic and abiotic stresses the missing link in understanding plant defence. Current Issues in Molecular Biology. 2017;**23**(1):1-16. DOI: 10.21775/cimb.023.001

[71] Di Serio F, De Stradis A, Delgado S, Flores R, Navarro B. Cytopathic effects incited by viroid RNAs and putative underlying mechanisms. Frontiers in

Plant Science. 2013;**3**:288. DOI: 10.3389/fpls.2012.00288

[72] Sasidharan R, Voesenek LA, Pierik R. Cell wall modifying proteins mediate plant acclimatization to biotic and abiotic stresses. Critical Reviews in Plant Sciences. 2011;**30**(6):548-562. DOI: 10.1080/07352689.2011.615706

[73] Sastry KS, Zitter TA. Plant Virus and Viroid Diseases in the Tropics. Vol. 2. Dordrecht: Springer; 2014. 489 p. DOI: 10.1007/978-94-007-6524-5

[74] Wei L, John Martin JJ, Zhang H, Zhang R, Cao H. Problems and prospects of improving abiotic stress tolerance and pathogen resistance of oil palm. Plants. 2021;**10**(12):2622. DOI: 10.3390/plants10122622

[75] Miller DE. Root systems in relation to stress tolerance. HortScience. 1986;**21**(4):963-970. DOI: 10.3389/fpls.2016.01584

[76] Yavas I, Jamal MA, Ul Din K, Ali S, Hussain S, Farooq M. Drought-induced changes in leaf morphology and anatomy: Overview, implications and perspectives. Polish Journal of Environmental Studies. 2024;**33**(2):1517-1530. DOI: 10.15244/pjoes/174476

[77] Hayat F, Li J, Iqbal S, Peng Y, Hong L, Balal RM, et al. A mini review of citrus rootstocks and their role in high-density orchards. Plants. 2022;**11**(21):2876. DOI: 10.3390/plants11212876

[78] Jones RAC. Future scenarios for plant virus pathogens as climate change progresses. Advances in Virus Research. 2016;**95**:87-147. DOI: 10.1016/bs.aivir.2016.02.004

[79] Kadioglu A, Terzi R, Saruhan N, Saglam A. Current advances in the investigation of leaf rolling caused by biotic and abiotic stress factors. Plant Science. 2012;**182**:42-48. DOI: 10.1016/j.plantsci.2011.01.013

[80] Yao T, Xie R, Zhou C, Wu X, Li D. Roles of brassinosteroids signaling in biotic and abiotic stresses. Journal of Agricultural and Food Chemistry. 2023;**71**(21):7947-7960. DOI: 10.1021/acs.jafc.2c07493

[81] Bashir MA, Silvestri C, Ahmad T, Hafiz IA, Abbasi NA, Manzoor A, et al. Osmotin: A cationic protein leads to improve biotic and abiotic stress tolerance in plants. Plants. 2020;**9**(8):992. DOI: 10.3390/plants9080992

[82] Li S, Wu ZG, Zhou Y, Dong ZF, Fei X, Zhou CY, et al. Changes in metabolism modulate induced by viroid infection in the orchid Dendrobium officinale. Virus Research. 2022;**308**:198626. DOI: 10.1016/j.virusres.2021.198626

[83] Joubert M, Van den Berg N, Theron J, Swart V. Transcriptomics advancement in the complex response of plants to viroid infection. International Journal of Molecular Sciences. 2022;**23**(14):7677. DOI: 10.3390/ijms23147677

[84] Fujibayashi M, Suzuki T, Sano T. Mechanism underlying potato spindle tuber viroid affecting tomato (*Solanum lycopersicum*): Loss of control over reactive oxygen species production. Journal of General Plant Pathology. 2021;**87**:226-235. DOI: 10.1007/s10327-021-01000-1

[85] Baker CJ, Smith JM, Yarberry AJ, Rice C. Induction of apoplast phenolics in pepper (*Capsicum annuum*) leaves in response to pathogenic bacteria. Physiological and Molecular Plant Pathology. 2020;**109**:101453. DOI: 10.1016/j.pmpp.2019.101453

[86] Kumar R, Lal MK, Tiwari RK, Chourasia KN, Kumar A, Kumar R, et al. Investigating the interplay between tomato leaf curl New Delhi virus infection, starch metabolism and antioxidant defence system in potato (*Solanum tuberosum* L.). Antioxidants. 2023;**12**(7):1447. DOI: 10.3390/antiox12071447

[87] Nath VS, Shrestha A, Awasthi P, Mishra AK, Kocábek T, Matoušek J, et al. Mapping the gene expression spectrum of mediator subunits in response to viroid infection in plants. International Journal of Molecular Sciences. 2020;**21**(7):2498. DOI: 10.3390/ijms21072498

[88] Baillo EH, Kimotho RN, Zhang Z, Xu P. Transcription factors associated with abiotic and biotic stress tolerance and their potential for crops improvement. Genes. 2019;**10**(10):771. DOI: 10.3390/genes10100771

[89] Wai AH, Naing AH, Lee DJ, Kim CK, Chung MY. Molecular genetic approaches for enhancing stress tolerance and fruit quality of tomato. Plant Biotechnology Reports. 2020;**14**:515-537. DOI: 10.1007/s11816-020-00638-1

[90] Makarova S, Makhotenko A, Spechenkova N, Love AJ, Kalinina NO, Taliansky M. Interactive responses of potato (*Solanum tuberosum* L.) plants to heat stress and infection with potato virus Y. Frontiers in Microbiology. 2018;**9**:2582. DOI: 10.3389/fmicb.2018.02582

[91] Arif Y, Singh P, Siddiqui H, Bajguz A, Hayat S. Salinity induced physiological and biochemical changes in plants: An omic approach towards salt stress tolerance. Plant Physiology and Biochemistry. 2020;**156**:64-77. DOI: 10.1016/j.plaphy.2020.08.042

[92] Kumar K, Chakraborty S. Roles of long non-coding RNAs in plant virus interactions. Journal of Plant Biochemistry and Biotechnology. 2021;**30**(4):684-697. DOI: 10.1007/s13562-021-00697-7

[93] Shah WH, Rasool A, Saleem S, Mushtaq NU, Tahir I, Hakeem KR, et al. Understanding the integrated pathways and mechanisms of transporters, protein kinases, and transcription factors in plants under salt stress. International Journal of Genomics. 2021;**2021**:5578727. DOI: 10.1155/2021/5578727

[94] Sofy AR, Hmed AA, MA AEL-A, Dawoud RA, Elshaarawy RF, Sofy MR. Mitigating effects of bean yellow mosaic virus infection in faba bean using new carboxymethyl chitosan-titania nanobiocomposites. International Journal of Biological Macromolecules. 2020;**163**:1261-1275. DOI: 10.1016/j.ijbiomac.2020.07.066

[95] Prasad PVV, Staggenborg SA, Ristic Z. Impacts of drought and/or heat stress on physiological, developmental, growth, and yield processes of crop plants. In: Aroca R, editor. Plant Responses to Drought Stress: From Morphological to Molecular Features. Berlin: Springer; 2012. pp. 301-355. DOI: 10.1007/978-3-642-32653-0_13

[96] Bhattacharya A. Effect of low temperature stress on photosynthesis and allied traits: A review. In: Wani SH, Herath H, editors. Physiological Processes in Plants under Low Temperature Stress. Singapore: Springer; 2022. DOI: 10.1007/978-981-16-9037-2_3

[97] Matoušek J, Orctová L, Steger G, Škopek J, Moors M, Dědič P, et al. Analysis of thermal stress-mediated PSTVd variation and biolistic inoculation of progeny of viroid "thermomutants"

to tomato and Brassica species. Virology. 2004;**323**(1):9-23. DOI: 10.1016/j.virol.2004.02.010

[98] Moshe A, Pfannstiel J, Brotman Y, Kolot M, Sobol I, Czosnek H, et al. Stress responses to tomato yellow leaf curl virus (TYLCV) infection of resistant and susceptible tomato plants are different. Metabolomics. 2012;**6**(S1):2153-0769. DOI: 10.4172/2153-0769.S1-006

[99] Viggiano L, de Pinto MC. Dynamic DNA methylation patterns in stress response. In: Rajewsky N, Jurga S, Barciszewski J, editors. Plant Epigenetics. Cham: Springer; 2017. pp. 281-302. DOI: 10.1007/978-3-319-55520-1_15

[100] Afroz A, Ali GM, Mir A, Komatsu S. Application of proteomics to investigate stress-induced proteins for improvement in crop protection. Plant Cell Reports. 2011;**30**:745-763. DOI: 10.1007/s00299-010-0982-x

[101] Sano T. Progress in 50 years of viroid research—Molecular structure, pathogenicity, and host adaptation. Proceedings of the Japan Academy. Series B, Physical and Biological Sciences. 2021;**97**(7):371-401. DOI: 10.2183/pjab.97.020

[102] Di Serio F, Izadpanah K, Hajizadeh M, Navarro B. Viroids infecting the grapevine. In: Meng B, Martelli G, Golino D, Fuchs M, editors. Grapevine Viruses: Molecular Biology, Diagnostics and Management. Cham: Springer; 2017. pp. 373-392. DOI: 10.1007/978-3-319-57706-7_19

[103] Lipiec J, Doussan C, Nosalewicz A, Kondracka K. Effect of drought and heat stresses on plant growth and yield: A review. International Agrophysics. 2013;**27**(4):463-477. DOI: 10.2478/intag-2013-0017

[104] Khalid MF, Hussain S, Ahmad S, Ejaz S, Zakir I, Ali MA, et al. Impacts of abiotic stresses on growth and development of plants. In: Wani SH, Herath H, editors. Plant Tolerance to Environmental Stress. Boca Raton: CRC Press; 2019. pp. 1-8. DOI: 10.1201/9780203705315

[105] Mirzayeva S, Huseynova I, Özmen CY, Ergül A. Physiology and gene expression analysis of tomato (*Solanum lycopersicum* L.) exposed to combined-virus and drought stresses. Plant Pathology Journal. 2023;**39**(5):466-478. DOI: 10.5423/PPJ.OA.07.2023.0103

[106] Mihaljević S, Milanović J, Oklestkova J, Novák O. The role of phytohormones in plant-viroid interactions. In: Çiftçi YO, editor. Applied Plant Biotechnology for Improving Resistance to Biotic Stress. Cambridge: Academic Press; 2020. pp. 321-342. DOI: 10.1016/B978-0-12-816030-5.00015-X

[107] Shteinberg M, Mishra R, Anfoka G, Altaleb M, Brotman Y, Moshelion M, et al. Tomato yellow leaf curl virus (TYLCV) promotes plant tolerance to drought. Cells. 2021;**10**(11):2875. DOI: 10.3390/cells10112875

[108] Zribi I, Ghorbel M, Brini F. Pathogenesis related proteins (PRs): From cellular mechanisms to plant defense. Current Protein & Peptide Science. 2021;**22**(5):396-412. DOI: 10.2174/1389203721999201231212736

[109] Hasanuzzaman M, Hakeem K, Nahar K, Alharby H. Plant Abiotic Stress Tolerance. Cham, Switzerland: Springer; 2019. p. 490. DOI: 10.1007/978-3-030-06118-0

[110] Majeed A, Muhammad Z. Salinity: A major agricultural

problem—Causes, impacts on crop productivity and management strategies. In: Hasanuzzaman M, Hakeem K, Nahar K, Alharby H, editors. Plant Abiotic Stress Tolerance. Cham, Switzerland: Springer; 2019. pp. 83-100. DOI: 10.1007/978-3-030-06118-0_3

[111] Jiang T, Zhou T. Unraveling the mechanisms of virus-induced symptom development in plants. Plants (Basel). 2023;**12**(15):2830. DOI: 10.3390/plants12152830

[112] Pan LL, Miao H, Wang Q, Walling LL, Liu SS. Virus-induced phytohormone dynamics and their effects on plant-insect interactions. The New Phytologist. 2021;**230**(4):1305-1320. DOI: 10.1111/nph.17261

[113] Rahman A, Sinha KV, Sopory SK, Sanan-Mishra N. Influence of virus-host interactions on plant response to abiotic stress. Plant Cell Reports. 2021;**40**(11):2225-2245. DOI: 10.1007/s00299-021-02718-0

[114] Li Z, Xu X, Leng X, He M, Wang J, Cheng S, et al. Roles of reactive oxygen species in cell signaling pathways and immune responses to viral infections. Archives of Virology. 2017;**162**(3):603-610. DOI: 10.1007/s00705-016-3130-2

[115] Mishra R, Shteinberg M, Shkolnik D, Anfoka G, Czosnek H, Gorovits R. Interplay between abiotic (drought) and biotic (virus) stresses in tomato plants. Molecular Plant Pathology. 2022;**23**(4):475-488. DOI: 10.1111/mpp.13172

[116] Abbasi H, Jamil M, Haq A, Ali S, Ahmad R, Malik Z, et al. Salt stress manifestation on plants, mechanism of salt tolerance and potassium role in alleviating it: A review. Zemdirbyste-Agriculture. 2016;**103**(2):229-238. DOI: 10.13080/z-a.2016.103.030

[117] Shahid MA, Sarkhosh A, Khan N, Balal RM, Ali S, Rossi L, et al. Insights into the physiological and biochemical impacts of salt stress on plant growth and development. Agronomy. 2020;**10**(7):938. DOI: 10.3390/agronomy10070938

[118] Parida AK, Das AB. Salt tolerance and salinity effects on plants: A review. Ecotoxicology and Environmental Safety. 2005;**60**(3):324-349. DOI: 10.1016/j.ecoenv.2004.06.010

[119] Acosta-Motos JR, Ortuño MF, Bernal-Vicente A, Diaz-Vivancos P, Sanchez-Blanco MJ, Hernandez JA. Plant responses to salt stress: Adaptive mechanisms. Agronomy. 2017;**7**(1):18. DOI: 10.3390/agronomy7010018

[120] Bai Y, Kissoudis C, Yan Z, Visser RGF, van der Linden G. Plant behaviour under combined stress: Tomato responses to combined salinity and pathogen stress. The Plant Journal. 2018;**93**(4):781-793. DOI: 10.1111/tpj.13800

[121] Omari Alzahrani F. Metabolic engineering of osmoprotectants to elucidate the mechanism (s) of salt stress tolerance in crop plants. Planta. 2021:**253**(1):24. DOI: 10.1007/s00425-020-03550-8

[122] Zandi P, Schnug E. Reactive oxygen species, antioxidant responses and implications from a microbial modulation perspective. Biology. 2022;**11**(2):155. DOI: 10.3390/biology11020155

[123] Poudel M, Mendes R, Costa LAS, Bueno CG, Meng Y, Folimonova SY, et al. The role of plant-associated bacteria, fungi, and viruses in drought stress mitigation. Frontiers in Microbiology. 2021;**12**:743512. DOI: 10.3389/fmicb.2021.743512

[124] Ramegowda V, Mysore KS, Senthil-Kumar M. Virus-induced gene silencing is a versatile tool for unraveling the functional relevance of multiple abiotic-stress-responsive genes in crop plants. Frontiers in Plant Science. 2014;**5**:323. DOI: 10.3389/fpls.2014.00323

[125] Prasad A, Sett S, Prasad M. Plant-virus-abiotic stress interactions: A complex interplay. Environmental and Experimental Botany. 2022;**199**:104869. DOI: 10.1016/j.envexpbot.2022.104869

[126] Li T, Huang Y, Xu ZS, Wang F, Xiong AS. Salicylic acid-induced differential resistance to the tomato yellow leaf curl virus among resistant and susceptible tomato cultivars. BMC Plant Biology. 2019;**19**:1-14. DOI: 10.1186/s12870-019-1784-0

[127] Choudhury S, Hu H, Larkin P, Meinke H, Shabala S, Ahmed I, et al. Agronomical, biochemical and histological response of resistant and susceptible wheat and barley under BYDV stress. PeerJ. 2018:**6**:e4833. DOI: 10.7717/peerj.4833

[128] Kumar K, Kumar M, Kim SR, Ryu H, Cho YG. Insights into genomics of salt stress response in rice. Rice (N Y). 2013;**6**(1):27. DOI: 10.1186/1939-8433-6-27

[129] Uzma IH, Ghori Z, Ali SH, Sheikh S, Gul A. Wheat responses to stress and biotechnological approaches for improvement. In: Hasanuzzaman M, Nahar K, Hossain M, editors. Singapore: Springer; 2019. DOI: 10.1007/978-981-13-6883-7_14

[130] Al-Askar AA, Aseel DG, El-Gendi H, Sobhy S, Samy MA, Hamdy E, et al. Antiviral activity of biosynthesized silver nanoparticles from pomegranate (*Punica granatum* L.) peel extract against tobacco mosaic virus. Plants. 2023;**12**(11):2103. DOI: 10.3390/plants12112103

[131] Parvathi MS, Antony PD, Kutty MS. Multiple stressors in vegetable production: Insights for trait-based crop improvement in cucurbits. Frontiers in Plant Science. 2022;**13**:861637. DOI: 10.3389/fpls.2022.861637

[132] Rai MK, Shekhawat NS. Recent advances in genetic engineering for improvement of fruit crops. Plant Cell Tissue and Organ Culture. 2014;**116**:1-15. DOI: 10.1007/s11240-013-0389-9

[133] Gupta P, Sharma R, Sharma MK, Sharma MP, Satpute GK, Singla-Pareek SL, et al. Signaling crosstalk between biotic and abiotic stress responses in soybean. In: Miransari M, editor. Abiotic and Biotic Stresses in Soybean Production. London: Academic Press; 2016: pp. 27-52. DOI: 10.1016/B978-0-12-801536-0.00002-5

[134] Cueto-Ginzo IA, Serrano L, Sin E, Rodríguez R, Morales JG, Lade SB, et al. Exogenous salicylic acid treatment delays initial infection and counteracts alterations induced by *Maize dwarf mosaic virus* in the maize proteome. Physiological and Molecular Plant Pathology. 2016:**96**:47-59. DOI: 10.1016/j.pmpp.2016.07.001

[135] Seleiman MF, Al-Suhaibani N, Ali N, Akmal M, Alotaibi M, Refay Y, et al. Drought stress impacts on plants and different approaches to alleviate its adverse effects. Plants. 2021;**10**:259. DOI: 10.3390/plants10020259

[136] Fang Y, Xiong L. General mechanisms of drought response and their application in drought resistance improvement in plants. Cellular and Molecular Life Sciences. 2015;**72**:673-689. DOI: 10.1007/s00018-014-1767-0

[137] Ejaz S, Fahad S, Anjum MA, Nawaz A, Naz S, Hussain S, et al. Role of osmolytes in the mechanisms of antioxidant defense of plants. In: Fahad S, Saeed MA, Khan MS, editors. Sustainable Agriculture Reviews. Vol. 39. Cham: Springer; 2020. pp. 95-117. DOI: 10.1007/978-3-030-38881-2_4

[138] Huang B. Role of root morphological and physiological characteristics in drought resistance of plants. In: Wilkinson RE, editor. Plant-Environment Interactions. 2nd ed. Boca Raton: CRC Press; 2000. pp. 57-82. DOI: 10.1201/9780824746568.ch2

[139] Jiang S, Postovit L, Cattaneo A, Binder EB, Aitchison KJ. Epigenetic modifications in stress response genes associated with childhood trauma. Frontiers in Psychiatry. 2019;**10**:477343. DOI: 10.3389/fpsyt.2019.00808

[140] Pinheiro C, António C, Ortuño MF, Dobrev PI, Hartung W, Thomas-Oates J, et al. Initial water deficit effects on Lupinus albus photosynthetic performance, carbon metabolism, and hormonal balance: Metabolic reorganization prior to early stress responses. Journal of Experimental Botany. 2011;**62**(14):4965-4974. DOI: 10.1093/jxb/err194

[141] Pallas V, García JA. How do plant viruses induce disease? Interactions and interference with host components. The Journal of General Virology. 2011;**92**(Pt. 12):2691-2705. DOI: 10.1099/vir.0.034603-0

[142] Muthamilarasan M, Prasad M. Plant innate immunity: An updated insight into defense mechanism. Journal of Biosciences. 2013;**38**:433-449. DOI: 10.1007/s12038-013-9302-2

[143] Hipper C, Brault V, Ziegler-Graff V, Revers F. Viral and cellular factors involved in phloem transport of plant viruses. Frontiers in Plant Science. 2013;**4**:154. DOI: 10.3389/fpls.2013.00154

[144] Yang YX, Ahammed GJ, Wu C, Fan SY, Zhou YH. Crosstalk among jasmonate, salicylate and ethylene signaling pathways in plant disease and immune responses. Current Protein & Peptide Science. 2015;**16**(5):450-461. DOI: 10.2174/13892037 16666150330141638

[145] Zvereva AS, Pooggin MM. Silencing and innate immunity in plant defense against viral and non-viral pathogens. Viruses. 2012;**4**(11):2578-2597. DOI: 10.3390/v4112578

[146] Liang J, Deng G, Long H, Pan Z, Wang C, Cai P, et al. Virus-induced silencing of genes encoding LEA protein in Tibetan hulless barley (*Hordeum vulgare* ssp. vulgare) and their relationship to drought tolerance. Molecular Breeding. 2012;**30**:441-451. DOI: 10.1007/s11032-011-9633-3

[147] Mirzayeva S, Huseynova I, Özmen CY, Ergül A. Physiology and gene expression analysis of tomato (*Solanum lycopersicum* L.) exposed to combined-virus and drought stresses. Plant Pathology Journal. 2023;**39**:5, 466. DOI: 10.5423/PPJ.OA.07.2023.0103

[148] Chen LJ, Ren H, Deng XG, Li YN, Cha WQ, Lin HH, et al. Effects of light intensity on the susceptibility of *Nicotiana tabacum* to cucumber mosaic virus. Journal of General Plant Pathology. 2015;**81**:399-408. DOI: 10.1007/s10327-015-0602-2

[149] Pasev G, Kostova D, Sofkova S. Identification of genes for resistance to bean common mosaic virus and bean common mosaic necrosis virus in snap bean (*Phaseolus vulgaris* L.) breeding

lines using conventional and molecular methods. Journal of Phytopathology. 2014;**162**(1):19-25. DOI: 10.1111/jph.12149

[150] Kalatskaja JN, Baliuk NV, Rybinskaya KI, Herasimovich KM, Yalouskaya NA, Yarullina LG, et al. Increasing potato sustainability to PVY under water deficiency by Bacillus bacteria with salicylic acid and methyl jasmonate. International Journal of Plant Biology. 2023;**14**(1):312-328. DOI: 10.3390/ijpb14010026

[151] Radwan DEM, Fayez KA, Mahmoud SY, Hamad A, Lu G. Salicylic acid alleviates growth inhibition and oxidative stress caused by zucchini yellow mosaic virus infection in *Cucurbita pepo* leaves. Physiological and Molecular Plant Pathology. 2006;**69**(4-6):172-181. DOI: 10.1016/j.pmpp.2007.04.004

[152] Encabo JR, Macalalad-Cabral RJA, Matres JMK, Coronejo SCTP, Jonson GB, Kishima Y, et al. Infection with an asymptomatic virus in rice results in a delayed drought response. Functional Plant Biology. 2020;**47**(3):239-249. DOI: 10.1071/FP19241

[153] Źróbek-Sokolnik A. Temperature stress and responses of plants. In: Ahmad P, Prasad M, editors. Environmental Adaptations and Stress Tolerance of Plants in the Era of Climate Change. New York: Springer; 2012. pp. 113-134. DOI: 10.1007/978-1-4614-0815-4_5

[154] Niu Y, Xiang Y. An overview of biomembrane functions in plant responses to high-temperature stress. Frontiers in Plant Science. 2018;**9**:915. DOI: 10.3389/fpls.2018.00915

[155] Belhadj Slimen I, Najar T, Ghram A, Dabbebi H, Ben Mrad M, Abdrabbah M. Reactive oxygen species, heat stress and oxidative-induced mitochondrial damage. A review. International Journal of Hyperthermia. 2014;**30**(7):513-523. DOI: 10.3109/02656736.2014.971446

[156] Prakash V, Sharma V, Devendran R, Prajapati R, Ahmad B, Kumar R. A transition from enemies to allies: How viruses improve drought resilience in plants. Stress Biology. 2024;**4**(1):33. DOI: 10.1007/s44154-024-00172-y

[157] Lamalakshmi Devi E, Kumar S, Basanta Singh T, Sharma SK, Beemrote A, Devi CP, et al. Adaptation strategies and defence mechanisms of plants during environmental stress. In: Ghorbanpour M, Varma A, editors. Medicinal Plants and Environmental Challenges. Cham: Springer; 2017. pp. 359-413. DOI: 10.1007/978-3-319-68717-9_20

[158] Khan Z, Shahwar D. Role of heat shock proteins (HSPs) and heat stress tolerance in crop plants. In: Roychowdhury R, Choudhury S, Hasanuzzaman M, Srivastava S, editors. Sustainable Agriculture in the Era of Climate Change. Cham: Springer; 2020. pp. 211-234. DOI: 10.1007/978-3-030-45669-6_9

[159] Li T, Xu X, Li Y, Wang H, Li Z, Li Z. Comparative transcriptome analysis reveals differential transcription in heat-susceptible and heat-tolerant pepper (*Capsicum annum* L.) cultivars under heat stress. Journal of Plant Biology. 2015;**58**:411-424. DOI: 10.1007/s12374-015-0423-z

[160] Anfoka G, Moshe A, Fridman L, Amrani L, Rotem OR, Kolot M, et al. Tomato yellow leaf curl virus infection mitigates the heat stress response of plants grown at high temperatures. Scientific Reports. 2016;**6**(1):19715. DOI: 10.1038/srep19715

[161] Soosaar JL, Burch-Smith TM, Dinesh-Kumar SP. Mechanisms of plant resistance to viruses. Nature Reviews. Microbiology. 2005;**3**(10):789-798. DOI: 10.1038/nrmicro1239

[162] Paudel DB, Sanfaçon H. Exploring the diversity of mechanisms associated with plant tolerance to virus infection. Frontiers in Plant Science. 2018;**9**:410882. DOI: 10.3389/fpls.2018.01575

[163] Grimmer MK, Foulkes MJ, Paveley ND. Foliar pathogenesis and plant water relations: A review. Journal of Experimental Botany. 2012;**63**(12):4321-4331. DOI: 10.1093/jxb/ers143

[164] Dixit S, Sivalingam PN, Baskaran RM, Senthil-Kumar M, Ghosh PK. Plant responses to concurrent abiotic and biotic stress: Unravelling physiological and morphological mechanisms. Plant Physiology Reports. 2024;**29**(1):6-17. DOI: 10.1007/s40502-023-00766-0

[165] Ul Haq S, Khan A, Ali M, Khattak AM, Gai WX, Zhang HX, et al. Heat shock proteins: Dynamic biomolecules to counter plant biotic and abiotic stresses. International Journal of Molecular Sciences. 2019;**20**(21):5321. DOI: 10.3390/ijms20215321

Chapter 4

Perspective Chapter: The Double Edge Sword – Modulations of Inflammatory Responses – Parasite Survival Strategy or Host Tolerance Mechanisms

Diana Ahu Prah, Henrietta Mensah-Brown, Yaw Aniweh, Evans Nii and Anna A. Kafintu-Kwashie

Abstract

Parasite organisms, including *Plasmodium*, *Toxoplasma*, *Leishmania*, and *Helminths*, rely on the host for survival, causing considerable morbidity and mortality in endemic regions. It is generally acknowledged that the clinical symptoms associated with parasitic infection are driven in part by exuberant pro-inflammatory cytokines produced when the organism interacts with the host innate immune system. Consequently, the host has evolved immune regulatory responses to limit excessive pro-inflammatory responses in order to tolerate the damage caused by the parasite or avoid immune-mediated pathology. However, suppressing pro-inflammatory cytokines appears to be detrimental to parasite removal, promoting pathogen persistence. Notably, parasites have also evolved strategies for exploiting the host's inflammatory responses to limit immune elimination and cause chronic infections. Thus, modulation of the host inflammatory responses results in a permissive environment, protecting the host against immune-mediated tissue injury while simultaneously allowing the parasite to survive, complete its life cycle, and spread the infections. Host immune modifications following natural infections can be particularly challenging for vaccination efforts, especially in endemic regions. If the immune system of infrequently infected individuals adapts to frequent antigenic stimulation, with the immune cells becoming hypo responsive to further antigen stimulation, then even vaccination may fail to achieve the necessary protection.

Keywords: host-parasite interaction, pro-inflammatory, anti-inflammatory, modulations, survival

1. Introduction

Parasitic infections are very common worldwide, particularly in the developing countries. Despite advances in research and therapeutic options, parasitic infections still cause serious illnesses like malaria as well as other frequently neglected tropical diseases such as helminthiasis, leishmaniasis, and trypanosomiasis. Parasitic diseases do not only afflict humans but also many domestic and wild animals, having a tremendous impact on underdeveloped countries and societies [1]. Parasites include organisms that live in or on another living thing (a host), benefiting from the host for their survival, development, reproduction, and propagation [1, 2]. Three main types of parasites have been defined to cause diseases in humans: protozoa, *Helminths* (endoparasites that reside inside the host), and ectoparasites such as lice, mosquitoes, bugs, ticks, and fleas that feed on the blood or other products of the host [3].

Parasitic organisms can inflict damage to several host tissues. For example, studies in *Helminths* have demonstrated that larval migration into the lungs is accompanied by an intense lungs and airways infiltration of inflammatory cells, severe intra-alveolar hemorrhage, and hyperactivity [4], and this can comprise normal respiratory function. The *Plasmodium* parasites reside in the red blood cells; hence, in principle, the parasite-infected red blood cells can spread to several organs and tissues. Subsequently, the parasite has been noted to cause several pathological conditions, such as kidney failure, severe anemia, acidosis, respiratory distress, bone loss, and cerebral malaria [5].

Accordingly, to ensure survival, the host must defend itself against parasites and the damage inflicted. Upon parasitic infection, the host immune system activates multiple anti-parasitic responses to eliminate and resist the parasites [3, 6]. Both the innate and adaptive arms of the immune system are activated with the subsequent production of multiple effector molecules, including cytokines and chemokines. Typically, pro-inflammatory cytokines such as interleukin-1β (IL-1β), tumor necrosis factor-alpha (TNF-α), IL-6, and interferon-gamma (IFN-γ) are secreted very early as components of the host's anti-parasitic mechanisms by dendritic cells, monocytes or macrophages, natural killer cells, and activated T cells. These cytokines induce multiple protective effects, which are beneficial through limiting parasite burden (**Figure 1**) [7–9]. For example, during *Entamoeba histolytica* infection, IFN-γ produced by natural kill cells protects against amebic liver abscess [10]. Likewise, TNF-α protects against parasite burden and infection-associated pathology during *Trypanosome* infections [11]. The inflammatory response pathways and mechanisms have been described in detail elsewhere [9, 12]. At the site of injury, pro-inflammatory cytokines can recruit and activate neutrophils, monocytes, and macrophages and induce acute responses to repair the injury [13, 14]. In some cases, the host immune responses can neutralize the parasites with no trace that the parasites were ever there in the host. However, under certain conditions, the parasites can persist for a long time without causing obvious clinical symptoms, with the infection becoming chronic (this phenomenon is generally described as an asymptomatic infection). The vast majority of parasite infections go undiagnosed and untreated, which can significantly impact infection transmission.

It has become common knowledge that host pro-inflammatory responses in an attempt to destruct and eliminate the parasites can also act to worsen the infection. For example, in *Plasmodium* infection, IFN-γ in combination with IL-12 and TNF-α induces phagocytosis and promotes increased clearance of parasitized red blood cells [15], however elevated levels are closely related to the severity of malaria and anemia [16–18]. In the

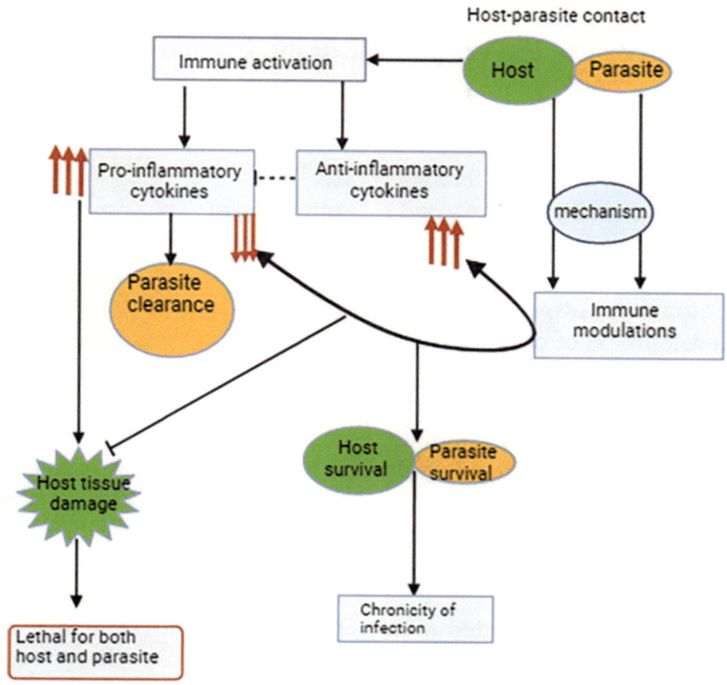

Figure 1.
The host parasites play a critical role in modulating inflammatory responses, allowing infection to persist over time. In the early phase of host-parasite contact, pro-inflammatory cytokines are induced and contribute to parasite clearance. Excessive pro-inflammatory activity promotes host tissue damage, which can be lethal, and in the course of the host death, the parasite may also not survive. Anti-inflammatory play an important role in limiting pro-inflammatory responses and preventing tissue damage. To ensure survival, both the host and parasite induce immune regulatory mechanisms that limit the production of pro-inflammatory activity and promote the increased production of anti-inflammatory responses, which can help control tissue damage.

event of imbalance and dysregulation, excessive pro-inflammatory cytokine production results in oxidative stress and an inflammatory burst, which in turn cause collateral tissue damage and severe disease in parasitic infection (**Figure 1**) [19–21]. Higher levels of pro-inflammatory cytokines enhance the expression of adhesion molecules on endothelial cells, facilitating the increased migration and entrapment of leucocytes into the microvasculature of organs and thereby disrupting normal function [22, 23]. This phenomenon, whereby the host responses induced against invading pathogens also increase the severity of the infection, is referred to as immunopathology [24].

Given the potential detrimental effects of exuberant pro-inflammatory cytokines, it makes sense that the hosts will carefully modulate and regulate these powerful inflammatory responses to guarantee survival. Consequently, in the course of the infection, anti-inflammatory cytokines such as IL-10 and transforming growth factor beta (TGF-β) are produced to regulate the activity of pro-inflammatory cytokines [9, 25]. Mice lacking the IL-0 gene succumb to infection with *Toxoplasma gondii* in the early phase of the infection as a result of the inability to control excessive pro-inflammatory responses [26, 27]. Further, following repeated parasite infections, transcriptional and epigenetic changes that dampen pro-inflammatory gene expressions are activated in the innate immune cells [28]. Individuals who are protected from severe *Plasmodium* infection pathologies are generally characterized by a shift from increased production of pro-inflammatory cytokines to predominately anti-inflammatory responses [29]. Moreover, regulatory T cells (Tregs) with immunosuppressive activity [30] and which attenuate pro-inflammatory responses are expanded [31] in chronic parasitic infections. Thus, modulating the production and activity of pro-inflammatory cytokines serves as a protective mechanism for the host, reducing tissue damage and the severity of clinical disease. This mechanism is recognized as a vital strategy for host survival.

Although attenuation of exacerbated pro-inflammatory responses is protective against several clinical manifestations of parasitic infection, it might not directly influence pathogen fitness. Indeed, there is strong evidence that modulation in the host pro-inflammatory responses results in inefficient parasite clearance and contributes to the development of chronic infection [32–34]. Thus, despite ongoing potent immune surveillance, some parasites can escape the immune system and persist in the host for long periods, with the infection progressing to the chronic phase. Thus, in the context of pro-inflammatory modulations, the host is tolerating the pathogen and avoiding severe outcomes of the infection, and this phenomenon has been termed as disease tolerance [35, 36]. Literature suggests that mammals have evolved to coexist in a symbiotic relationship with various species of pathogens [37]. It has emerged that adequate, or even more importantly, balanced pro-inflammatory and anti-inflammatory responses are key for the host tissues to defend against external invasive parasites and internal tissue injury. Considering the effectiveness of inflammatory regulatory networks in preventing tissue and organ damage while also allowing parasite persistence, the question arises: are these regulatory mechanisms controlled by the host or the parasite?

The ultimate goal of parasites from an evolutionary perspective is to survive and pass on the infection to a new host. In this context, it will be a selective advantage for the parasite not to overwhelm the host; the infected host has to survive so that the parasites can complete their life cycle and spread the infection. Present scientific knowledge shows that parasites strive to use several mechanisms to exploit the host inflammatory mechanisms for their own survival. The parasitic secreted molecule HpARI (H. polygyrus alarmin release inhibitor) provides a good example of how a parasitic organism hijacks the host pro-inflammatory cytokine responses for their benefit. Numerous innate immune cells express the IL-33/ receptor ST2L, which strongly induces the production of cytokines. A number of studies have highlighted the significance of this signaling pathway in both the clearance of parasites and the aggravation of inflammation in allergic illnesses [38, 39]. It has been demonstrated that the *Helminth* parasite *Heligmosomoides polygyrus* releases HpARI, which binds to active IL-33, prevents it from interacting with immune cell receptors, and thereby lessens response [40]. By limiting the action of IL-33, the parasitic worm avoids elimination and seizes the chance to propagate and increase its burden [39, 40]. Additionally, the host is also spare from excessive inflammation and tissue damage [41]. When this prolongs the host's life, the infectious period is likely to increase, and therefore the parasite prevalence will also increase [42]. Thus, as a result of this "harmonious symbiosis," several parasites can establish long-term infection in the host. Under the context of chronic infection, the parasite can take advantage to learn the host inflammatory responses generated against it and possibly adapt strategies to evade these responses. It is thus tempting to suggest that the parasite is driving these inflammatory-regulating mechanisms to ensure host survival, allowing the parasite to persist and increasing the probability of transmission [43].

2. Modifications of inflammatory pathways: Host disease tolerance and surviving strategy

Hosts that are infected with parasitic organisms must mount a powerful immune response to eradicate the parasites and prevent mortality. Inflammation is an essential immune system's response against invading pathogens, controls tissue injuries, and also maintains homeostasis [12]. It has, however, emerged that many of the clinical symptoms associated with parasitic infection are caused by the over-production of pro-inflammatory cytokines secreted when the organism interacts with the host immune system [21, 44, 45]. Interestingly, epidemiological data from endemic regions of the world suggests that humans have developed mutualistic relationships with parasitic organisms, including *Helminths* and *Plasmodium*. Reports indicate that the majority of *Plasmodium* and *Helminth* infections fail to elicit clinical signs of illness and remain asymptomatic, which can persist for months or years [46]. The absence of clinical manifestations indicates that humans have developed effective strategies to defend themselves against severe manifestations of this infection [31, 47].

According to growing scientific evidence, following repeated parasitic infections, multiple changes in the host immune systems, including epigenetic and transcriptional reprogramming of innate immune cells, expansion of regulatory immune subsets, and metabolic adaptations, are induced toward a reduction in pro-inflammatory cytokine production [48–50]. Functional changes, including reduced pro-inflammatory cytokines production to increased anti-inflammatory cytokines production, have been reported in individuals' chronically infected with Plasmodium parasites [29]. This suppression of pro-inflammatory responses represents an emerging concept to explain why repeated exposure to most parasite organisms results in silent infection without clinical signs.

Notably, enhancing the function and numbers of regulatory immune cells with immunosuppressive activity is crucial to controlling excessive pro-inflammatory responses while still allowing the host to fight off infections. Regulatory B cells (Bregs), regulatory T cells (Tregs), and regulatory natural killer cells have been identified as key components of the immune network functioning during chronically asymptomatic infections [51]. Besides their roles in antibody secreting and antigen presentation functions, B cells have also been confirmed as cytokine producers [52]. The study by Amu et al. investigating the role of Bregs in managing allergic airway inflammation in a murine model found that Bregs can prevent and reverse allergic airway inflammation through interactions with FoxP3-positive Tregs [53]. The study specifically emphasizes that Bregs generate IL-10, which aids in immune response modulation and inflammation reduction in allergic situations [53]. Bregs with functional production of IL-10 have been identified to expand rapidly during parasitic infection with *Leishmania major* [54]. The study also showed Bregs downregulated IL-12 production [54]. Bregs from mice infected with *Plasmodium* parasites expanded rapidly in the course of the infection, secreted IL-10, and inhibited the production of IFN-γ [55]. In addition, the study showed that adoptive transfer of IL-10-producing Bregs led to enhanced survival of recipient mice following infection with a lethal parasite [55]. Thus, IL-10 produced by Bregs helps to mitigate the severity of cerebral malaria [56]. Similarly, Bao et al. demonstrated that Breg cells are the primary source of plasma IL-10 during experimental cerebral malaria, and transferring these cells to naïve mice provided protection against mortality [57]. Furthermore, an increased proportion of Bregs that predominantly produced IL-10 was observed in *Toxoplasma gondii* infection [51]. Like parasitic infection, the frequency of IL-10-producing Bregs has also been shown to increase in the course of bacterial infection [58]. It has become clear that Tregs, which conditionally express co-inhibitory receptors, predominantly produce IL-10 and other anti-inflammatory cytokines and dampen pro-inflammatory responses, are expanded over the course of repeated parasitic infections [31]. Altogether, the expansion of these regulatory cells with anti-inflammatory activity during parasitic infections may indicate a mechanism by which the host adapts to avoid excessive pro-inflammatory-associated pathology [53, 59].

Innate defenses triggered by monocytes/macrophages and dendritic cells play a crucial role in immune responses against intracellular pathogens [8]. A deleterious role of monocytes/macrophages through the increased production of pro-inflammatory cytokines has been reported in human parasitic diseases [60]. Subsequently, in partially protected individuals poised to control malaria fever, monocytes have more regulatory properties and produce lower levels of inflammatory cytokines than monocytes from non-immune individuals [49]. Thus, as a host disease tolerance and survival strategy to parasitic infection, monocytes have transitioned from the inflammatory states to the regulatory states with reduced production of pro-inflammatory cytokines and increased production of anti-inflammatory cytokines [49]. Epigenetic modifications have been observed in the innate immune cells of individuals living in malaria-endemic settings, and these alterations have been correlated with decreased pro-inflammatory cytokine production [49, 61]. Interestingly, these modifications at the epigenetic level were associated with the phenotype of asymptomatic infection (clinically immune) and predicted long-term future disease tolerance [28].

Activated Vγ9Vδ2 (γδ2) T lymphocytes help to control intracellular parasitic burden [62] but are also major producers of pro-inflammatory cytokines [63]. Data so far indicate that protection against clinical malaria is correlated with loss of peripheral Vδ2+ cells and with the remaining Vδ2+ subset of γδ T cells having an increased expression of immunoregulatory genes with a reduced capacity to produce pro-inflammatory cytokines [64, 65]. The host, by limiting the expansion of these cells and regulating their activities [64], may represent an adapted strategy to avoid pathology.

Helminth infections are extremely prevalent and are linked to serious health issues as well as increased co-morbidities such as anemia, pulmonary fibrosis, and digestive diseases [66]. The fact that *Helminths* can survive in the human host for decades with the host rarely developing severe complications associated with the infection suggests that humans have evolved effective defense mechanisms against these parasites with persistent exposure to parasite agents, the host inflammatory responses pathway are modified such that responses to subsequent or persistent infection may either be diminished or enhanced compared to the primary responses [67, 68]. Individuals can become tolerant to the inflammatory effects of parasitic infection.

Notably, the modulation of inflammatory pathways occurs downstream from pathogen sensing, sometimes at the hematopoietic stem and progenitor level [28]. These modifications and reprogramming gradually develop across multiple years and through repeated exposure. Furthermore, the majority of these modifications are maintained in the absence of parasitic infections. It therefore seems probable that the host actively adopts these regulatory signatures in order to control parasite burden without succumbing to the damaging effects of exacerbated inflammation or tissue damage.

3. Manipulation of host inflammatory responses: The best asset for parasite survival

Immune cells protect against parasites by producing cytokines. And as emphasized above, excessive pro-inflammatory responses result in severe pathology and death, which is harmful to both the parasite and the host, as the host's death will also halt the infection (**Figure 1**). In this regard, the invading parasite will gain a selective advantage by limiting lethal inflammatory responses, which might allow it to maintain the health of the host. When this allows the host to live longer, the infectious period is likely to increase, as does parasite prevalence. Parasites are known to establish chronic infections in their host, an indication of their ability to evade and alter host defenses to propagate. Given the effectiveness of the host inflammatory responses to intracellular pathogens, it can be suspected that the parasites will hijack these responses for their own benefits [43]. Indeed, several studies have demonstrated that, as an escape mechanism, parasites secrete a range of products that strongly interfere with the inflammatory responses or influence certain types of immune cells (**Table 1**) [87, 88]. For example, monocyte locomotion inhibitory factor (MLIF) is an immunosuppressive molecule produced by *Entamoeba histolytica* that limits pro-inflammatory response and also inhibits nitric oxide production [81].

Helminths infections are generally well tolerated and rather prevalent in natural populations. *Helminths*' ability to persist in the human host for decades is assumed to be due to the induction of immunoregulatory mechanisms, resulting in an anti-inflammatory environment [39, 88–90]. For example, IL-33 bound to the host receptor (ST2, IL1RL1, and IL-33R) induces strong cytokine production, which provides protection against parasites but has also been linked to several pathological conditions like asthma and inflammatory bowel disease [91–93]. HpARI2 produced by *Heligmosomoides polygyrus* binds to IL-33 and inhibits activation of the IL-33/receptor signaling, thus accounting for its inflammation-suppressing properties [40, 69]. Inhibition of the IL-33/receptor signaling pathway increases parasite load [40] while sparing the host from several immediate damages [41, 92]. Another evidence for the exploitation of the inflammatory responses by parasitic organisms is the production of the TGF-β mimic protein TGM by the parasitic *Helminth H. polygyrus*. TGM, although structurally unrelated to the mammalian TGF-β or other members of the TGF-β family, binds strongly to mammalian plasma membrane TGF-β receptors and perfectly replicates TGF-β's functional activity [70, 71, 94].

During chronic infection, *Toxoplasma gondii* and its derived products increase IL-10 production [95, 96]. Increased levels of IL-10 were found to suppress the production of pro-inflammatory cytokines such as IL-12, IFN-γ, and TNF-α, and IL-6, favoring anti-inflammatory responses [96]. Under this condition, the host is protected from lethal inflammation, ensuring survival and, eventually, the survival of the parasite species [96, 97]. *Leishmania* parasites promote the increased production of the Host Immune Inhibitory Ligand CD200, which inhibits

Immunomodulatory molecule	Examples of parasites	Mechanism involved	References
Alarmin release inhibitor (HpARI)	*Heligmosomoides polygyrus*	Inhibit human and mouse IL-33.	[40, 69]
TGF-β mimic (TGM)	*Heligmosomoides polygyrus*	Induction of Treg cells.	[70, 71]
Macrophage Migration Inhibitory Factor homolog-1 (MIF-1)	*Anisakis simplex*	Interacts with Toll-like receptor 2 (TLR2), leading to a reduction in inflammatory responses associated with allergic airway diseases.	[72]
Anti-inflammatory protein-2 (AIP-2)	*Necator americanus*	Suppresses proliferation of T cells and also induces Treg responses that suppress experimental asthma.	[73]
Abundant Larval Transcript (ALT)	*Brugia malayi*	Upregulates GATA-3 and SOCS-1 transcripts, the inhibitor of IFN-γ signaling.	[74]
leucyl0 aminopeptidase (ES-62)	*Brugia malayi*	Impair dendritic function, diminished production of both pro-inflammatory cytokines and type I interferon.	[75]
IL-4 inducing principle from *S. mansoni eggs* (IPSE/alpha-1)	*Schistosoma mansoni*	Bind to basophils and induce the release of the anti-inflammatory cytokines IL-4 and 1 L-13.	[76]
Haemonchus contortus excretory/secretory proteins (HcESPs)	*Haemonchus contortus*	Downregulate the TLR4/NF-κB/MAPKs signaling pathway. Activate the STAT3/PD-L1 pathway and promote the activity of regulatory T cells (Tregs).	[77]
cyclooxygenase (COX)-like enzyme (EhPGE2)	*Entamoeba histolytica*	Synthesizes EhPGE2, which inhibits macrophage th1 cytokine release, respiratory burst, and nitric oxide production.	[78–80]
Monocyte locomotion inhibitory factor (MLIF)	*Entamoeba histolytica*	Anti-inflammatory activities by inhibiting NO production.	[81]
Plasmodium berghei T-cell immunomodulatory protein (PbTIP)	*Plasmodium spp.*	Exhibits binding on the surface of macrophages and reduces their inflammatory cytokine response while upregulating the anti-inflammatory cytokines such as TGF-β and IL-10.	[82]
Hemozoin	*Plasmodium spp.*	Maintains immunosuppressive activity by inhibiting dendritic cell maturation, differentiation, and function.	[83–85]
Em492-antigen	*Echinococcus multilocularis*	Contributing to immunosuppressive events that occur at the host-parasite interface.	[32]
Trichinella spiralis (TS) 53-kDa protein (rTsP53)	*Trichinella spiralis*	Reduce production of pro-inflammatory cytokines and increase regulatory T-cell responses.	[86]

Table 1.
Examples of parasites derived immunomodulatory molecules with the potential to create a permissive environment.

macrophage activation and hence limits inflammation [98]. Likewise, *Leishmania* exosomes have been demonstrated to be predominately anti-inflammatory [99]. EgTeg is an anti-inflammatory molecule produced by *Echinococcus granulosus*. These parasite-derived immunomodulatory molecules contribute to the persistence of the infection.

Under chronic conditions, when the host immune system has little impact on the parasites, the parasites are likely to learn the host defenses and devise ways to evolve into a new strain that can evade immunological destruction. When *Trypanosomes* multiply in the host bloodstream, the host generates an immune response that is only effective against the population of *Trypanosomes* that express a specific variant surface glycoprotein (VSG). Trypanosomes that have transitioned to a different VSG coat prevent destruction [100]. A similar mechanism has been observed in *P. falciparum* infection. The parasite-expressed variant surface antigens, particularly PfEMP1 (*P falciparum* erythrocyte membrane protein), are crucial to *P. falciparum* pathogenesis because they allow the infected erythrocyte to adhere to endothelial cells and avoid clearance by the spleen's filtering function [101]. Clearly, the host anti-PfEMP1-specific antibody plays a crucial role in parasite clearance during infection [102]. PfEMP1 is encoded by ~60 members of the var. gene, although only one is expressed at a time. The parasite shifts to a new type to escape host responses to the previously expressed one. Such periodic antigenic variation is critical to parasites' ability to initiate and sustain long-term chronic infection [101].

4. Impact of host immune modulation and vaccination responses

Vaccination is one of the most effective public health measures used to protect humans against infectious disease. A major question that remains is the effect of host immune epigenetic reprogramming and resultant inflammatory modulations on responses to vaccines against both parasitic and non-parasitic infections. If the immune system of infrequently infected individuals adapts to frequent antigenic stimulation, with the immune cells becoming hypo responsive to further antigen stimulation, then even vaccination may fail to achieve the necessary protection.

Recent studies show that BCG vaccination causes epigenetic reprogramming that reduces the production of pro-inflammatory cytokines such as IL-1β while increasing the production of anti-inflammatory cytokines such as IL-37 [103]. There is some evidence, both in human and animal studies, that acute *Helminth* infections impaired vaccine efficacy against seasonal influenza [104, 105]. More concerningly, reduced vaccine responses were observed even after the infection was cleared. Further to the above, tetanus toxoid and hepatitis B vaccine responses were found to decline rapidly in individuals with schistosomiasis [106]. It has been known from the days of malaria therapy that sterile immunity against malaria can be achieved in naïve subjects [107, 108], however immunity in individuals residing in malaria-endemic areas has largely been suggested to be rather poor [109]. A recent study examining COVID-19 immune responses in malaria-exposed individuals demonstrated that individuals with high anti-Pf antibodies showed an attenuated T-cell response, resulting in a low cytokine production [110]. Taken together, this may suggest that host immune modulation following natural infection with a parasitic pathogen has a profound impact on the host responses to vaccination in individuals with previous high exposure.

5. Conclusions

The development of effective vaccines against parasites remains a major challenge. Though there is currently a vaccine against malaria, low efficacy remains a major challenge. In this review, we examined immunomodulation/tolerance due to frequent exposure and the impact it has on vaccination to parasitic and non-parasitic diseases and the development of anti-parasitic vaccines. It is clear that there are more questions than answers. More data is needed so we can better understand the development of immunity and the impact that frequent exposure to diseases such as malaria has on vaccination.

Funding

DAP is supported by a Postdoctoral fellowship from WACCBIP-DELTAS II Advanced Technology Training fellowship.

Conflict of interest

The authors declare no conflicts of interest.

Author details

Diana Ahu Prah[1*], Henrietta Mensah-Brown[1], Yaw Aniweh[1], Evans Nii[1] and Anna A. Kafintu-Kwashie[2]

1 West African Centre for Cell Biology of Infectious Pathogens, University of Ghana, Legon, Ghana,

2 Department of Medical Laboratory Sciences, School of Biomedical and Allied Health Sciences, University of Ghana, Korle Bu, Accra, Ghana

*Address all correspondence to: daprah@gmail.com

References

[1] Cummings RD, Hokke CH, Haslam SM. Parasitic Infections. In: Varki A, Cummings RD, Esko JD, Stanley P, Hart GW, Aebi M, et al, editors. Essentials of Glycobiology [Internet]. 4th ed. Cold Spring Harbor (NY): Cold Spring Harbor Laboratory Press; 2022. Chapter 43

[2] Atique R, Saeed HA, Haidar A, Sharif J, Naveed A, Ijaz A, et al. Host-parasite interactions; from co-evolutionary changes to genomic insights. Global Journal of Universal Studies. 2024;**1**(1):88-107

[3] Chulanetra M, Chaicumpa W. Revisiting the mechanisms of immune evasion employed by human parasites. Frontiers in Cellular and Infection Microbiology. 2021;**11**:702125

[4] Enobe CS, Araujo C, Perini A, Martins M, Macedo M, MFD M-S. Early stages of *Ascaris suum* induce airway inflammation and hyperreactivity in a mouse model. Parasite Immunology. 2006;**28**(9):453-461

[5] Coban C, Lee MSJ, Ishii KJ. Tissue-specific immunopathology during malaria infection. Nature Reviews Immunology. 2018;**18**(4):266-278

[6] Onyilagha C, Uzonna JE. Host immune responses and immune evasion strategies in African trypanosomiasis. Frontiers in Immunology. 2019;**10**:2738

[7] Tanaka T, Nagasawa H, Fujisaki K, Suzuki N, Mikami T. Growth-inhibitory effects of interferon-γ on *Neospora caninum* in murine macrophages by a nitric oxide mechanism. Parasitology Research. 2000;**86**:768-771

[8] da Silva SC, Boaventura V, Cardoso CR, Tavares N, Lordelo MJ, Noronha A, et al. CD8+ granzyme B+–mediated tissue injury vs. CD4+ IFNγ+–mediated parasite killing in human cutaneous leishmaniasis. Journal of Investigative Dermatology. 2013;**133**(6):1533-1540

[9] Obeagu EI. Role of cytokines in immunomodulation during malaria clearance. Annals of Medicine and Surgery. 2024;**86**(5):2873-2882

[10] Lotter H, González-Roldán N, Lindner B, Winau F, Isibasi A, Moreno-Lafont M, et al. Natural killer T cells activated by a lipopeptidophosphoglycan from Entamoeba histolytica are critically important to control amebic liver abscess. PLoS Pathogens. 2009;**5**(5):e1000434

[11] Magez S, Radwanska M, Beschin A, Sekikawa K, De Baetselier P. Tumor necrosis factor alpha is a key mediator in the regulation of experimental *Trypanosoma brucei* infections. Infection and Immunity. 1999;**67**(6):3128-3132

[12] Chen L, Deng H, Cui H, Fang J, Zuo Z, Deng J, et al. Inflammatory responses and inflammation-associated diseases in organs. Oncotarget. 2018;**9**(6):7204

[13] Xiao T, Yan Z, Xiao S, Xia Y. Proinflammatory cytokines regulate epidermal stem cells in wound epithelialization. Stem Cell Research and Therapy. 2020;**11**(1):232

[14] Roth M, Nauck M, Tamm M, Perruchoud AP, Ziesche R, Block L. Intracellular interleukin 6 mediates platelet-derived growth factor-induced proliferation of nontransformed cells. National Academy of Sciences of the United States of America. 1995;**92**(5):1312-1316

[15] Oyegue-Liabagui SL, Bouopda-Tuedom AG, Kouna LC, Maghendji-Nzondo S, Nzoughe H, Tchitoula-Makaya N, et al. Pro-and anti-inflammatory cytokines in children with malaria in Franceville, Gabon. American Journal of Clinical and Experimental Immunology. 2017;**6**(2):9

[16] Mandala WL, Msefula CL, Gondwe EN, Drayson MT, Molyneux ME, MacLennan CA. Cytokine profiles in Malawian children presenting with uncomplicated malaria, severe malarial anemia, and cerebral malaria. Clinical and Vaccine Immunology. 2017;**24**(4):e00533-e00516

[17] Ong'echa JM, Davenport GC, Vulule JM, Hittner JB, Perkins DJ. Identification of inflammatory biomarkers for pediatric malarial anemia severity using novel statistical methods. Infection and Immunity. 2011;**79**(11):4674-4680

[18] Mahittikorn A, Mala W, Masangkay FR, Kotepui KU, Wilairatana P, Kotepui M. Increased interferon-γ levels and risk of severe malaria: A meta-analysis. Scientific Reports. 2022;**12**(1):18917

[19] Andrade BB, Reis-Filho A, Souza-Neto SM, Clarêncio J, Camargo LM, Barral A, et al. Severe plasmodium vivax malaria exhibits marked inflammatory imbalance. Malaria Journal. 2010;**9**:1-8

[20] Cruz LA, Barral-Netto M, Andrade BB. Distinct inflammatory profile underlies pathological increases in creatinine levels associated with *Plasmodium vivax* malaria clinical severity. PLoS Neglected Tropical Diseases. 2018;**12**(3):e0006306

[21] Saldanha MG, Pagliari C, Queiroz A, Machado PRL, Carvalho L, Scott P, et al. Tissue damage in human cutaneous leishmaniasis: Correlations between inflammatory cells and molecule expression. Frontiers in Cellular and Infection Microbiology. 2020;**10**:355

[22] del Pozo MA, Sanchez-Mateos P, Nieto M, Sanchez-Madrid F. Chemokines regulate cellular polarization and adhesion receptor redistribution during lymphocyte interaction with endothelium and extracellular matrix. Involvement of cAMP signaling pathway. The Journal of Cell Biology. 1995;**131**(2):495-508

[23] Li R, Ye JJ, Gan L, Zhang M, Sun D, Li Y, et al. Traumatic inflammatory response: Pathophysiological role and clinical value of cytokines. European Journal of Trauma and Emergency Surgery. Aug 2024;**50**(4):1313-1330

[24] Elliott DE, Weinstock JV. Helminth–host immunological interactions: Prevention and control of immune-mediated diseases. Annals of the New York Academy of Sciences. 2012;**1247**(1):83-96

[25] Deckert-Schlüter M, Buck C, Weiner D, Kaefer N, Rang A, Hof H, et al. Interleukin-10 downregulates the intracerebral immune response in chronic toxoplasma encephalitis. Journal of Neuroimmunology. 1997;**76**(1-2):167-176

[26] Suzuki Y, Sher A, Yap G, Park D, Neyer LE, Liesenfeld O, et al. IL-10 is required for prevention of necrosis in the small intestine and mortality in both genetically resistant BALB/c and susceptible C57BL/6 mice following peroral infection with toxoplasma gondii. The Journal of Immunology. 2000;**164**(10):5375-5382

[27] Gazzinelli RT, Wysocka M, Hieny S, Scharton-Kersten T, Cheever A, Kühn R, et al. In the absence of endogenous IL-10, mice acutely infected with toxoplasma

gondii succumb to a lethal immune response dependent on CD4+ T cells and accompanied by overproduction of IL-12, IFN-gamma and TNF-alpha. Journal of Immunology (Baltimore, Md: 1950). 1996;**157**(2):798-805

[28] Nideffer J, Ty M, Donato M, John R, Kajubi R, Ji X, et al. Clinical immunity to malaria involves epigenetic reprogramming of innate immune cells. PNAS Nexus. 2024;**3**(8):pgae325

[29] Portugal S, Moebius J, Skinner J, Doumbo S, Doumtabe D, Kone Y, et al. Exposure-dependent control of malaria-induced inflammation in children. PLoS Pathogens. 2014;**10**(4):e1004079

[30] Dikiy S, Rudensky AY. Principles of regulatory T cell function. Immunity. 2023;**56**(2):240-255

[31] Boyle MJ, Engwerda CR, Jagannathan P. The impact of plasmodium-driven immunoregulatory networks on immunity to malaria. Nature Reviews Immunology. Sep 2024;**24**(9):637-653

[32] Jangpatarapongsa K, Chootong P, Sattabongkot J, Chotivanich K, Sirichaisinthop J, Tungpradabkul S, et al. Plasmodium vivax parasites alter the balance of myeloid and plasmacytoid dendritic cells and the induction of regulatory T cells. European Journal of Immunology. 2008;**38**(10):2697-2705

[33] Saeftel M, Volkmann L, Korten S, Brattig N, Al-Qaoud K, Fleischer B, et al. Lack of interferon-γ confers impaired neutrophil granulocyte function and imparts prolonged survival of adult filarial worms in murine filariasis. Microbes and Infection. 2001;**3**(3):203-213

[34] Kumar R, Singh N, Gautam S, Singh OP, Gidwani K, Rai M, et al. Leishmania specific CD4 T cells release IFNγ that limits parasite replication in patients with visceral leishmaniasis. PLoS Neglected Tropical Diseases. 2014;**8**(10):e3198

[35] McCarville J, Ayres J. Disease tolerance: Concept and mechanisms. Current Opinion in Immunology. 2018;**50**:88-93

[36] Medzhitov R, Schneider DS, Soares MP. Disease tolerance as a defense strategy. Science. 2012;**335**(6071):936-941

[37] Lozupone CA, Stombaugh JI, Gordon JI, Jansson JK, Knight R. Diversity, stability and resilience of the human gut microbiota. Nature. 2012;**489**(7415):220-230

[38] Chan BC, Lam CW, Tam L-S, Wong CK. IL33: Roles in allergic inflammation and therapeutic perspectives. Frontiers in Immunology. 2019;**10**:364

[39] Wiedemann M, Voehringer D. Immunomodulation and immune escape strategies of gastrointestinal helminths and schistosomes. Frontiers in Immunology. 2020;**11**:572865

[40] Osbourn M, Soares DC, Vacca F, Cohen ES, Scott IC, Gregory WF, et al. HpARI protein secreted by a helminth parasite suppresses interleukin-33. Immunity. 2017;**47**(4):739-751. e5

[41] McSorley HJ, Blair NF, Smith KA, McKenzie AN, Maizels RM. Blockade of IL-33 release and suppression of type 2 innate lymphoid cell responses by helminth secreted products in airway allergy. Mucosal Immunology. 2014;**7**(5):1068-1078

[42] Roy B, Kirchner J. Evolutionary dynamics of pathogen resistance and tolerance. Evolution. 2000;**54**(1):51-63

[43] Aliberti J. Host persistence: Exploitation of anti-inflammatory pathways by toxoplasma gondii. Nature Reviews Immunology. 2005;**5**(2):162-170

[44] Hunt NH, Grau GE. Cytokines: Accelerators and brakes in the pathogenesis of cerebral malaria. Trends in Immunology. 2003;**24**(9):491-499

[45] Clark IA, Budd AC, Alleva LM, Cowden WB. Human malarial disease: A consequence of inflammatory cytokine release. Malaria Journal. 2006;**5**:1-32

[46] Prah DA, Laryea-Akrong E. Asymptomatic low-density *Plasmodium falciparum* infections: Parasites under the host's immune radar? The Journal of Infectious Diseases. 2024;**229**(6):1913-1918

[47] Tran TM, Jones MB, Ongoiba A, Bijker EM, Schats R, Venepally P, et al. Transcriptomic evidence for modulation of host inflammatory responses during febrile *Plasmodium falciparum* malaria. Scientific Reports. 2016;**6**(1):31291

[48] Saeed S, Quintin J, Kerstens HH, Rao NA, Aghajanirefah A, Matarese F, et al. Epigenetic programming of monocyte-to-macrophage differentiation and trained innate immunity. Science. 2014;**345**(6204):1251086

[49] Guha R, Mathioudaki A, Doumbo S, Doumtabe D, Skinner J, Arora G, et al. Plasmodium falciparum malaria drives epigenetic reprogramming of human monocytes toward a regulatory phenotype. PLoS Pathogens. 2021;**17**(4):e1009430

[50] Fiyouzi T, Pelaez-Prestel HF, Reyes-Manzanas R, Lafuente EM, Reche PA. Enhancing regulatory T cells to treat inflammatory and autoimmune diseases. International Journal of Molecular Sciences. 2023;**24**(9):7797

[51] Jeong Y-I, Hong S-H, Cho S-H, Park MY, Lee S-E. Induction of IL-10-producing regulatory B cells following toxoplasma gondii infection is important to the cyst formation. Biochemistry and Biophysics Reports. 2016;**7**:91-97

[52] Harris DP, Haynes L, Sayles PC, Duso DK, Eaton SM, Lepak NM, et al. Reciprocal regulation of polarized cytokine production by effector B and T cells. Nature Immunology. 2000;**1**(6):475-482

[53] Amu S, Saunders SP, Kronenberg M, Mangan NE, Atzberger A, Fallon PG. Regulatory B cells prevent and reverse allergic airway inflammation via FoxP3-positive T regulatory cells in a murine model. Journal of Allergy and Clinical Immunology. 2010;**125**(5):1114-1124. e8

[54] Ronet C, Torre YH-L, Revaz-Breton M, Mastelic B, Tacchini-Cottier F, Louis J, et al. Regulatory B cells shape the development of Th2 immune responses in BALB/c mice infected with Leishmania major through IL-10 production. The Journal of Immunology. 2010;**184**(2):886-894

[55] Kalkal M, Chauhan R, Thakur RS, Tiwari M, Pande V, Das J. IL-10 producing regulatory B cells mediated protection against murine malaria pathogenesis. Biology. 2022;**11**(5):669

[56] Liu Y, Chen Y, Li Z, Han Y, Sun Y, Wang Q, et al. Role of IL-10-producing regulatory B cells in control of cerebral malaria in *Plasmodium berghei* infected mice. European Journal of Immunology. 2013;**43**(11):2907-2918

[57] Bao LQ, Huy NT, Kikuchi M, Yanagi T, Senba M, Shuaibu MN, et al. CD19 (+) B cells confer protection against experimental cerebral malaria in semi-immune rodent model. PLoS One. 2013;**8**(5):e64836

[58] Jeong YI, Hong SH, Cho SH, Lee WJ, Lee SE. Induction of IL-10-Producing CD1dhighCD5+ Regulatory B Cells Following Babesia Microti-Infection. PLoS One. 2012;**7**(10):e46553

[59] Le Huu D, Matsushita T, Jin G, Hamaguchi Y, Hasegawa M, Takehara K, et al. Donor-derived regulatory B cells are important for suppression of murine sclerodermatous chronic graft-versus-host disease. Blood, The Journal of the American Society of Hematology. 2013;**121**(16):3274-3283

[60] Giudice A, Vendrame C, Bezerra C, Carvalho LP, Delavechia T, Carvalho EM, et al. Macrophages participate in host protection and the disease pathology associated with *Leishmania braziliensis* infection. BMC Infectious Diseases. 2012;**12**:1-9

[61] Nideffer J, Ty M, Donato M, John R, Kajubi R, Ji X, et al. Disease tolerance acquired through repeated plasmodium infection involves epigenetic reprogramming of innate immune cells. BioRxiv. 2023;**2023**(04):19.537546

[62] Junqueira C, Polidoro RB, Castro G, Absalon S, Liang Z, Sen Santara S, et al. γδ T cells suppress plasmodium falciparum blood-stage infection by direct killing and phagocytosis. Nature Immunology. 2021;**22**(3):347-357

[63] Goodier MR, Lundqvist C, Hammarström ML, Troye-Blomberg M, Langhorne J. Cytokine profiles for human Vγ9+ T cells stimulated by *Plasmodium falciparum*. Parasite Immunology. 1995;**17**(8):413-423

[64] Jagannathan P, Kim CC, Greenhouse B, Nankya F, Bowen K, Eccles-James I, et al. Loss and dysfunction of Vδ2+ γδ T cells are associated with clinical tolerance to malaria. Science Translational Medicine. 2014;**6**(251):251ra117-251ra117

[65] Jagannathan P, Lutwama F, Boyle MJ, Nankya F, Farrington LA, McIntyre TI, et al. Vδ2+ T cell response to malaria correlates with protection from infection but is attenuated with repeated exposure. Scientific Reports. 2017;**7**(1):11487

[66] Oliveira FMS, da Paixão Matias PH, Kraemer L, Gazzinelli-Guimarães AC, Santos FV, Amorim CCO, et al. Comorbidity associated to *Ascaris suum* infection during pulmonary fibrosis exacerbates chronic lung and liver inflammation and dysfunction but not affect the parasite cycle in mice. PLoS Neglected Tropical Diseases. 2019;**13**(11):e0007896

[67] Netea MG, Domínguez-Andrés J, Barreiro LB, Chavakis T, Divangahi M, Fuchs E, et al. Defining trained immunity and its role in health and disease. Nature Reviews Immunology. 2020;**20**(6):375-388

[68] Netea MG, Joosten LA, Latz E, Mills KH, Natoli G, Stunnenberg HG, et al. Trained immunity: A program of innate immune memory in health and disease. Science. 2016;**352**(6284):aaf1098

[69] Jamwal A, Colomb F, McSorley HJ, Higgins MK. Structural basis for IL-33 recognition and its antagonism by the helminth effector protein HpARI2. Nature Communications. 2024;**15**(1):5226

[70] Johnston CJ, Smyth DJ, Kodali RB, White MP, Harcus Y, Filbey KJ, et al. A structurally distinct TGF-β mimic from an intestinal helminth parasite potently induces regulatory T cells. Nature Communications. 2017;**8**(1):1741

[71] Chauché C, Rasid O, Donachie AM, McManus CM, Löser S, Campion T, et al. Suppression of airway allergic eosinophilia by Hp-TGM, a helminth

mimic of TGF-β. Immunology. 2022;**167**(2):197-211

[72] Cho M, Park M, Kang S, Park S, Lyu J, Kim DH, et al. TLR 2-dependent amelioration of allergic airway inflammation by parasitic nematode type II MIF in mice. Parasite Immunology. 2015;**37**(4):180-191

[73] Navarro S, Pickering DA, Ferreira IB, Jones L, Ryan S, Troy S, et al. Hookworm recombinant protein promotes regulatory T cell responses that suppress experimental asthma. Science Translational Medicine. 2016;**8**(362):362ra143-362ra143

[74] Gomez-Escobar N, Bennett C, Prieto-Lafuente L, Aebischer T, Blackburn CC, Maizels RM. Heterologous expression of the filarial nematode alt gene products reveals their potential to inhibit immune function. BMC Biology. 2005;**3**:1-16

[75] Semnani RT, Sabzevari H, Iyer R, Nutman TB. Filarial antigens impair the function of human dendritic cells during differentiation. Infection and Immunity. 2001;**69**(9):5813-5822

[76] Knuhr K, Langhans K, Nyenhuis S, Viertmann K, Kildemoes AMO, Doenhoff MJ, et al. Schistosoma mansoni egg-released IPSE/alpha-1 dampens inflammatory cytokine responses via basophil interleukin (IL)-4 and IL-13. Frontiers in Immunology. 2018;**9**:2293

[77] Wen Z, Zhang Y, Feng J, Aimulajiang K, Aleem MT, Lu M, et al. Excretory/secretory proteins inhibit host immune responses by downregulating the TLR4/NF-κB/MAPKs signaling pathway: A possible mechanism of immune evasion in parasitic nematode *Haemonchus contortus*. Frontiers in Immunology. 2022;**13**:1013159

[78] Dey I, Keller K, Belley A, Chadee K. Identification and characterization of a cyclooxygenase-like enzyme from Entamoeba histolytica. National Academy of Sciences of the United States of America. 2003;**100**(23):13561-13566

[79] Lin J, Keller K, Chadee K. Entamoeba histolytica proteins modulate the respiratory burst potential by murine macrophages. Immunology. 1993;**78**(2):291

[80] Wang W, Chadee K. Entamoeba histolytica suppresses gamma interferon-induced macrophage class II major histocompatibility complex IA molecule and IA beta mRNA expression by a prostaglandin E2-dependent mechanism. Infection and Immunity. 1995;**63**(3):1089-1094

[81] Rico G, Leandro E, Rojas S, Giménez J, Kretschmer R. The monocyte locomotion inhibitory factor produced by Entamoeba histolytica inhibits induced nitric oxide production in human leukocytes. Parasitology Research. 2003;**90**:264-267

[82] Kalia I, Anand R, Quadiri A, Bhattacharya S, Sahoo B, Singh AP. Plasmodium berghei-released factor, Pb TIP, modulates the host innate immune responses. Frontiers in Immunology. 2021;**12**:699887

[83] Scorza T, Magez S, Brys L, De Baetselier P. Hemozoin is a key factor in the induction of malaria-associated immunosuppression. Parasite Immunology. 1999;**21**(11):545-554

[84] Skorokhod OA, Alessio M, Mordmuller B, Arese P, Schwarzer E. Hemozoin (malarial pigment) inhibits differentiation and maturation of human monocyte-derived dendritic cells: A peroxisome proliferator-activated receptor-γ-mediated effect. The Journal of Immunology. 2004;**173**(6):4066-4074

[85] Millington OR, Di Lorenzo C, Phillips RS, Garside P, Brewer JM. Suppression of adaptive immunity to heterologous antigens during plasmodium infection through hemozoin-induced failure of dendritic cell function. Journal of Biology. 2006;**5**:1-22

[86] Du L, Tang H, Ma Z, Xu J, Gao W, Chen J, et al. The protective effect of the recombinant 53-kDa protein of *Trichinella spiralis* on experimental colitis in mice. Digestive Diseases and Sciences. 2011;**56**:2810-2817

[87] Das K, Nozaki T. Parasite secretory molecules in host-parasite interaction. In: Eukaryome Impact on Human Intestine Homeostasis and Mucosal Immunology: Overview of the First Eukaryome Congress at Institut Pasteur. Paris; October 16-18, 2019. Springer International Publishing; 2020. pp. 193-212

[88] McSorley HJ, Hewitson JP, Maizels RM. Immunomodulation by helminth parasites: Defining mechanisms and mediators. International Journal for Parasitology. 2013;**43**(3-4):301-310

[89] Maizels RM, Yazdanbakhsh M. Immune regulation by helminth parasites: Cellular and molecular mechanisms. Nature Reviews Immunology. 2003;**3**(9):733-744

[90] Taylor MD, LeGoff L, Harris A, Malone E, Allen JE, Maizels RM. Removal of regulatory T cell activity reverses hyporesponsiveness and leads to filarial parasite clearance in vivo. The Journal of Immunology. 2005;**174**(8):4924-4933

[91] Liew FY, Girard J-P, Turnquist HR. Interleukin-33 in health and disease. Nature Reviews Immunology. 2016;**16**(11):676-689

[92] De Salvo C, Wang X-M, Pastorelli L, Mattioli B, Omenetti S, Buela KA, et al. IL-33 drives eosinophil infiltration and pathogenic type 2 helper T-cell immune responses leading to chronic experimental ileitis. The American Journal of Pathology. 2016;**186**(4):885-898

[93] Gieseck RL III, Wilson MS, Wynn TA. Type 2 immunity in tissue repair and fibrosis. Nature Reviews Immunology. 2018;**18**(1):62-76

[94] Redgrave RE, Singh E, Tual-Chalot S, Park C, Hall D, Bennaceur K, et al. Exogenous TGFβ1 and its mimic HpTGM attenuate the heart's inflammatory response to ischaemic injury and improve long term cardiac outcomes. bioRxiv. 2023;**2023**(04):18.537417

[95] Pestechian N, Shahreza HK, Faridnia R, Kalani H. Manipulation of IL-10 gene expression by toxoplasma gondii and its products. Medical Journal of the Islamic Republic of Iran. 2016;**30**:410

[96] Wilson EH, Wille-Reece U, Dzierszinski F, Hunter CA. A critical role for IL-10 in limiting inflammation during toxoplasmic encephalitis. Journal of Neuroimmunology. 2005;**165**(1-2):63-74

[97] Jeong Y-I, Hong S-H, Cho S-H, Lee W-J, Lee S-E. Toxoplasma gondii infection suppresses house dust mite extract-induced atopic dermatitis in NC/Nga mice. Allergy, Asthma and Immunology Research. 2015;**7**(6):557-564

[98] Vaine CA, Soberman RJ. The CD200–CD200R1 inhibitory signaling pathway: Immune regulation and host–pathogen interactions. Advances in Immunology. 2014;**121**:191-211

[99] Silverman JM, Clos J, Horakova E, Wang AY, Wiesgigl M, Kelly I, et al.

Leishmania exosomes modulate innate and adaptive immune responses through effects on monocytes and dendritic cells. The Journal of Immunology. 2010;**185**(9):5011-5022

[100] Morrison LJ, Steketee PC, Tettey MD, Matthews KR. Pathogenicity and virulence of African trypanosomes: From laboratory models to clinically relevant hosts. Virulence. 2023;**14**(1):2150445

[101] Nyarko PB, Claessens A. Understanding host–pathogen–vector interactions with chronic asymptomatic malaria infections. Trends in Parasitology. 2021;**37**(3):195-204

[102] Zhang X, Deitsch KW. The mystery of persistent, asymptomatic plasmodium falciparum infections. Current Opinion in Microbiology. 2022;**70**:102231

[103] Arts RJW, Moorlag S, Novakovic B, Li Y, Wang SY, Oosting M, et al. BCG vaccination protects against experimental viral infection in humans through the induction of cytokines associated with trained immunity. Cell Host and Microbe. 2018;**23**(1):89-100.e5

[104] Breloer M, Hartmann W. Filarial infections compromise influenza vaccination efficacy: Lessons from the mouse. Immunology Letters. 2023;**255**:62-66

[105] Hartmann W, Brunn M, Stetter N, Gagliani N, Muscate F, Stanelle-Bertram S. Helminth infections suppress the efficacy of vaccination against seasonal influenza. Cell Reports. 2019;**29**(8):2243-56 e4

[106] Riner DK, Ndombi EM, Carter JM, Omondi A, Kittur N, Kavere E, et al. Schistosoma mansoni infection can jeopardize the duration of protective levels of antibody responses to immunizations against hepatitis B and tetanus toxoid. PLoS Neglected Tropical Diseases. 2016;**10**(12):e0005180

[107] Roestenberg M, McCall M, Hopman J, Wiersma J, Luty AJ, van Gemert GJ, et al. Protection against a malaria challenge by sporozoite inoculation. New England Journal of Medicine. 2009;**361**(5):468-477

[108] Belnoue E, Costa F, Frankenberg T, Vigário AM, Voza T, Leroy N, et al. Protective T cell immunity against malaria liver stage after vaccination with live sporozoites under chloroquine treatment. The Journal of Immunology. 2004;**172**(4):2487-2495

[109] RTS SCTP. Efficacy and safety of the RTS, S/AS01 malaria vaccine during 18 months after vaccination: A phase 3 randomized, controlled trial in children and young infants at 11 African sites. PLoS Medicine. 2014;**11**(7):e1001685

[110] Tapela K, Prah DA, Tetteh B, Nuokpem F, Dosoo D, Coker A, et al. Cellular immune response to SARS-CoV-2 and clinical presentation in individuals exposed to endemic malaria. Cell Reports. 2024;**43**(8):114533

Chapter 5

The Dual Nature of Host-Parasite Interactions: Exploring Protozoans and Helminths in Symbiosis and Pathogenesis

Selorme Adukpo

Abstract

Symbiotic interactions between various organisms challenge the idea of parasitism, which mostly coronates a detrimental relationship between two organisms: the host, which does not benefit, and the other, the parasite, which benefits from the relationship, imposing a fitness cost and, in some cases, killing the host altogether. Further challenging the idea of parasitism is the idea of commensalism. This chapter delves into symbiosis, a ubiquitous long-term interaction between different species, from the mutualistic association between organisms of two distinct species, in which at least one of them benefits, and commensal, a relationship between two different species of organisms where one benefits from the relationship while the other one is neither harmed nor benefited, to parasitic relationships. The relationship between humans and two classes of parasites that are of medical importance, as they are capable of causing disease in humans are discussed. The chapter emphasizes the extent to which interactions between humans and these organisms can be symbiotic, providing information on the mode of transmission of the parasites to their human host and the possible benefits each class of organisms could derive from the relationship. The biological and evolutionary dynamics underlying these interactions are also addressed. The immunomodulatory effects of the so-called parasites and their potential health benefits to humans are discussed.

Keywords: helminths, symbiosis, protozoan parasites, symbiosis, mutualism, comensalism

1. Introduction

Protozoan parasites include several single-celled eukaryotes that exhibit various interactions, from mutualistic to parasitic exploitative associations with their hosts. These interactions greatly influence human health and biology, illustrating the complexity of the relationships between them and humans. This section provides information on protozoan parasites, their life cycles, pathogenic mechanisms, clinical manifestations of the disease, and the mutualistic and parasitic roles they play as they interact with humans. Protozoa are unicellular eukaryotic organisms, traditionally classified into various groups based on their movement mechanisms: amoeboid, flagellates, ciliates, and sporozoans. Protozoan parasites, including Sarcodina, Mastigophora, Ciliophora, Sporozoa, or Apicomplexa, infect humans through multiple pathways, including ingesting the infective stage in contaminated food or water, vector transmission, and direct contact.

1.1 Sarcodina (amoebas)

Amoebas belong to the Sarcodina group and are distinguished by temporary cytoplasmic extensions and pseudopodia, which are used for movement and food capture. These organisms inhabit aquatic environments and soil or exist as parasites in various hosts, including humans. They undergo asexual reproduction through binary fission and can form cysts to withstand harsh conditions, leading to survival in an unfavorable environment. Within this group are *Entamoeba histolytica*, *Naegleria fowleri*, and *Acanthamoeba* spp. [1].

1.1.1 Entamoeba histolytica

Entamoeba histolytica causes amoebiasis, a condition characterized by intestinal and extraintestinal infections. Once an individual ingests cysts from contaminated food or water sources, they transform into trophozoites in the intestine and infiltrate the mucosal lining of the large intestine as they establish a close relationship with the human host. In establishing a close relationship with humans, they cause abdominal discomfort, diarrhea, and, in severe cases, dysentery and amoebic liver abscesses can occur when the infection spreads to the liver and other organs, imposing a great cost to fitness on the host.

1.1.2 Acanthamoeba *spp.*

Acanthamoeba spp. are causative agents of granulomatous amoebic encephalitis, particularly in individuals with compromised immune systems, and *Acanthamoeba keratitis*, an eye infection often associated with contact lens usage [1].

1.1.3 Naegleria fowleri

Naegleria fowleri, commonly referred to as the "brain-eating amoeba," flourishes in warm water, including poorly maintained swimming pools or chlorinated water. Although a free-living organism, it infects humans when contaminated water enters the nasal passages, allowing the amoeba to migrate to the brain and feed on the brain cells to cause rare brain infections known as naegleriasis or primary amoebic meningoencephalitis that is often lethal [1].

1.1.4 Amoeba lifecycle and transmission

Amoebas typically show a life cycle alternating between an active feeding phase, trophozoite, and a dormant phase, cyst. In the case of humans, infection occurs through the ingestion of cysts from contaminated food or water. Upon reaching the host, the cysts release trophozoites to populate the large intestine (*Entamoeba histolytica*), various tissues and organs (*Acanthamoeba* spp), and eye (*Acanthamoeba keratitis*) to cause the associated diseases as indicated in the **Figure 1** [2].

1.2 Mastigophora (flagellates)

Flagellates, members of the Mastigophora group, are protozoans that utilize one or more whip-like structures called flagella for locomotion. This category encompasses both free-living and parasitic species that infect humans and other animals. Flagellates reproduce asexually through binary fission and are typically found in the blood of their host, the gastrointestinal tract, or tissues. Notable Mastigophora species include *Giardia lamblia, Trichomonas vaginalis*, among others.

1.2.1 Giardia lamblia

Giardia lamblia: This organism is responsible for giardiasis, a prevalent gastrointestinal infection characterized by diarrhea, abdominal cramps, and malabsorption. Giardia lamblia spreads through the ingestion of cysts from contaminated water, food, or direct person-to-person contact.

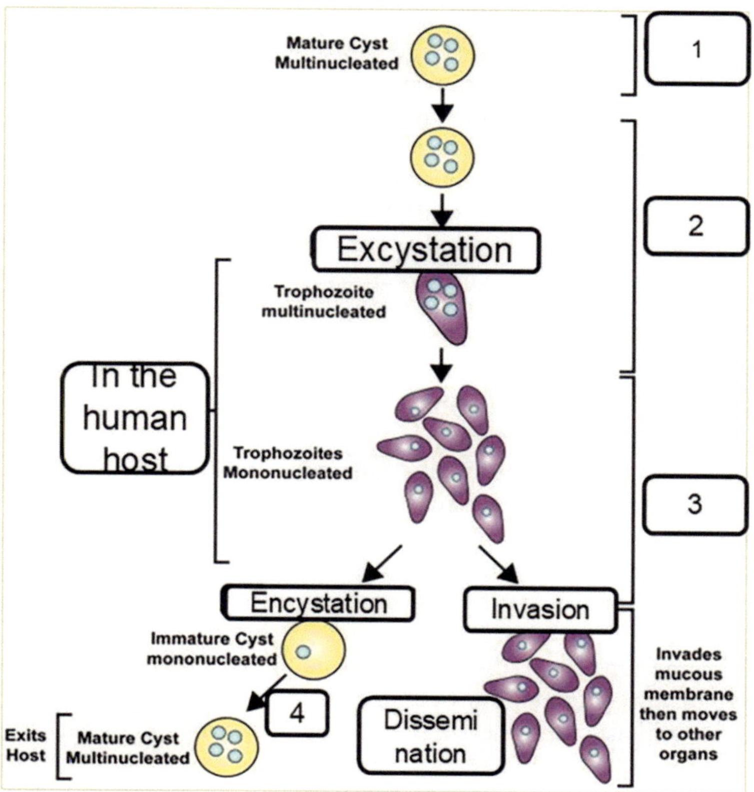

Figure 1.
Life cycle of pathogenic amoeba. (1) The mature cyst enters the human host through skin penetration, ingestion, or the lower respiratory tract via *inhalation. (2) The cyst moves to the small intestine where it undergoes excystation in the case of* Entamoeba histolytica*. (3) Release multiple trophozoites that migrate to the various tissues and organs, including the large intestine in the case of* Entamoeba histolytica*. Trophozoites may invade the mucous membrane of the host or encyst and exit the host for future infections.*

1.2.2 Trichomonas vaginalis

Trichomonas vaginalis: This species causes trichomoniasis, a sexually transmitted infection affecting the urogenital tract. Symptoms in women include vaginal discharge, itching, and discomfort during urination, while men are often asymptomatic carriers.

1.2.3 Leishmania *spp.*

Leishmania spp.: These cause leishmaniasis, a disease transmitted by sandflies. Depending on the species, leishmaniasis can present in three forms: cutaneous, mucocutaneous, or visceral. Visceral leishmaniasis, known as kala-azar, is the most severe form, affecting internal organs. Thus, the course of the disease depends on host susceptibility and the species of Leishmania involved: *L. major*, *L. amazonensis*, and *L. braziliensis* cause essentially mucocutaneous or cutaneous leishmaniasis, with the development of chronic ulcers on the skin or mucosa or self-healing. On the other hand, *Leishmania donovani* and *L. infantum* cause the visceral form of the disease, mainly affecting the liver, spleen, and bone marrow.

1.2.4 Trypanosoma *spp.*

Trypanosoma brucei: This species is responsible for African trypanosomiasis (sleeping sickness), a disease transmitted by the tsetse fly. The infection is characterized by intermittent fevers, headaches, and, in later stages, neurological symptoms such as sleep disturbances and confusion. On the other hand, Trypanosomiasis caused by *Trypanosoma brucei gambiense* and *T. brucei rhodesiense* are responsible for African Trypanomiasis. While T. *b. gambiense*, widely distributed on the West African continent is responsible for a chronic form of the disease, while *T. b. rhodesiense* found in central, eastern, and southern Africa, is responsible for the acute form of the disease. Altogether, *T. b. gambiense*, endemic in about 24 countries of West and Central Africa, causes a chronic disease that constitutes the majority of reported trypanosomiasis cases, with less than 10% of the cases are accounted for by *T. b. rhodesiense*, which is found in about 13 countries of eastern and southern Africa [3, 4].

The organism is vectored by a hematophagous insect, tsetse fly, and causes disease in vertebrate animals termed trypanosomiasis or sleeping sickness in humans and animal trypanosomiasis in animals. Once the tsetse fly picks up the protozoa during a blood meal, it undergoes various morphological changes in the intestine of the vector as it transitions from insects to the mammalian vertebrate host [5, 6].

Common symptoms of this disease include itchiness, headache, and joint pains, which slowly advance into numbness, poor coordination and trouble sleeping, seizures, coma, and ultimately death. Initial signs and symptoms appear a few weeks or months after infection. The disease develops rapidly with multi-organ and tissue invasion, including the brain.

Chagas disease is caused *Trypanosoma cruzi* is generally restricted to the Latin American continent but seems to be spreading to other areas like Europe and Africa. This species is transmitted by the feces of triatomine bugs (also known as "kissing bugs"). Chagas disease can lead to chronic cardiac and gastrointestinal complications.

1.2.5 Life cycle and transmission of flagellates

The lifecycle of flagellates often involves both direct and vector-mediated transmission. For example, *Giardia lamblia* is transmitted through ingesting cysts, while *Trypanosoma brucei* is transmitted through the bite of infected tsetse flies. Parasites often undergo developmental changes within the human host or vector. In the case of *Trypanosoma cruzi*, the parasite is transmitted through the feces of triatomine bugs. The parasites are deposited on the skin during a bite for a blood meal to develop the eggs of the triatomi bug. The biting site itches so when it is scratched, the parasite enters the host through mucous membranes or broken skin, leading to infection (**Figures 2** and **3**).

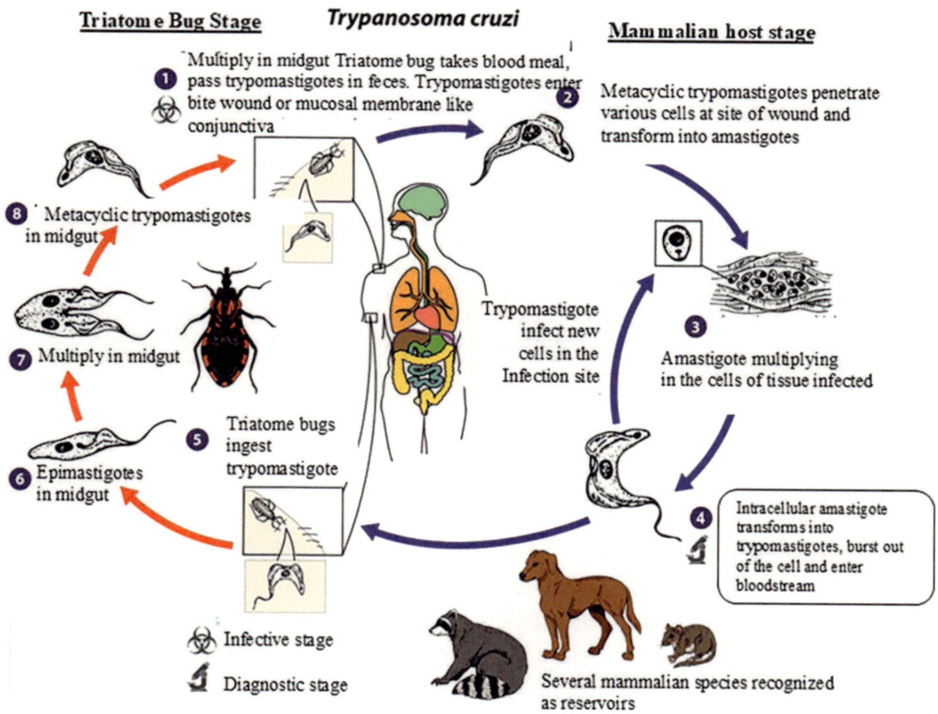

Figure 2.
Life cycle of Trypanosoma cruzi. *An infected insect vector takes bites for a blood meal, releasing the trypomastigote stage of the parasite in its feces. Trypomastigotes enter the host through the intact mucosal membrane or the bite wound. (2) In the host, the trypomastigotes invade cells and transform into distinct intracellular amastigotes. (3) The amastigotes multiply by binary fission. (4) Amastigotes transform into trypomastigotes. (5) Trypomastigotes are released into the bloodstream. (6) The "kissing" bug gets infected by feeding on humans or animals with infection. (7) The trypomastigotes differentiate into epimastigotes in the midgut of the vector. (8) The parasites multiply and differentiate in the midgut into infective metacyclic trypomastigotes ready to be injected into another human or animal [www.cdc.gov/parasites/and accessed on 31st October 2024].*

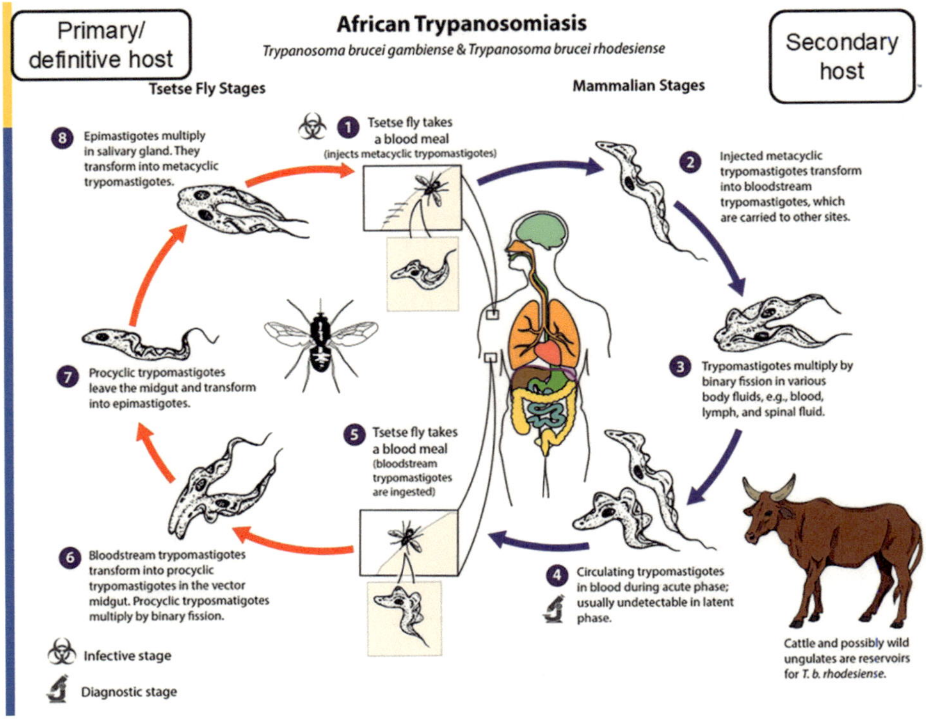

Figure 3.
*Tsetse fly (*Glossina *spp) takes bites for a blood meal, releasing the metacyclic trypomastigote stage of the parasite into the host. (2) Inside the host, they differentiate into bloodstream trypomastigotes. (3) They are transported to various sites throughout the body, including the lymph and spinal fluid. They replicate continuously by binary fission while being transported. (4/5) The fly becomes infected with bloodstream trypomastigotes upon taking a blood meal on an infected human host or mammal. (6) In the midgut of the tsetse fly, the parasites differentiate into procyclic trypomastigotes, multiplied by binary fission. (7) Procyclic trypomastigote leaves the midgut of the vector and transforms into epimastigotes. (8) The epimastigotes finally reach the salivary glands of the tsetse fly and multiply repeatedly by binary fission. In total, the life cycle in the fly lasts for about 3 weeks [www.cdc.gov/parasites/. Accessed on 31st October 2024].*

1.3 Ciliophora (ciliates)

Ciliophora (Ciliates): Ciliates are a group of protozoans with numerous hair-like structures called cilia that are used for movement and feeding. Most ciliates are free-living organisms found in aquatic environments, but a few species, including *Balantidium coli*, are parasitic and can infect humans. *Balantidium coli* is the only ciliate known to cause infection or balantidiasis, a gastrointestinal infection in farmers and rural inhabitants. Symptoms include abdominal pain, diarrhea, and dysentery when the case is severe. *Balantidium coli* is typically transmitted through cysts in contaminated water or food [7], often associated with close contact with pigs, the reservoir host. *Balantidium coli* undergoes a two-stage lifecycle: the trophozoite (active and reproductive) and cyst (dormant and infectious) phases. Human infection occurs through the consumption of cysts in contaminated food or water. Upon entering the host, cysts release trophozoites that establish themselves in the large intestine. This infection is prevalent in areas with inadequate sanitation and among individuals, including farmers and rural inhabitants, who frequently contact pigs (**Figure 4**) [7].

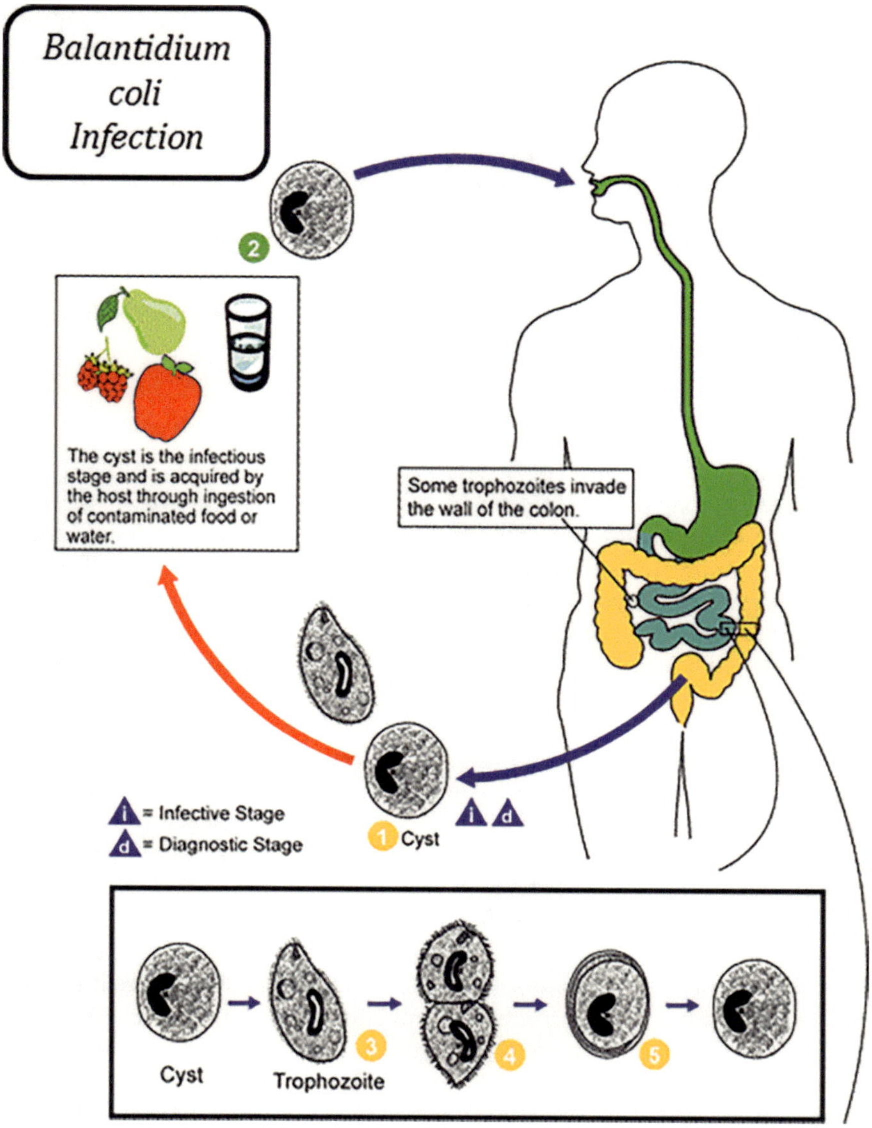

Figure 4.
Life cycle of Balantidium coli *infection in humans. (1) The trophozoites and cysts released in feces, and (2) Cysts-contaminated drinking water, food, fruits, and vegetables, and the infection may spread to other humans. The bottom panel shows the pattern of encystment and asexual reproduction [7].*

1.4 Sporozoa or Apicomplexa

Sporozoa or Apicomplexa are obligate intracellular parasitic protozoans. They include *Plasmodium* spp., *Toxoplasma gondii*, *Cryptosporidium* spp., and *Babesia* spp. Members of these protozoan groups lack noticeable locomotory structures like cilia or flagella, so they utilize a specialized apical complex to penetrate host cells. These organisms reproduce both sexually and asexually, often with intricate lifecycles involving multiple hosts. For instance, the life cycle of *P. falciparum, P. ovale, P. malaria, P. vivax,* and *P. knowlesi* encompass both human and mosquito hosts. In humans, parasites undergo asexual reproduction in the liver and red blood cells, while sexual reproduction occurs in the mosquito vector.

1.4.1 Plasmodium *spp.*

Plasmodium spp.: These parasites cause malaria, a globally significant parasitic disease. Transmission occurs through the bite of infected female Anopheles mosquitoes, which introduce the parasites into the bloodstream. Plasmodium infects red blood cells in the human host, resulting in symptoms such as fever, chills, and anemia. Severe cases can be fatal or lead to organ failure. Five Plasmodium species cause human malaria, with *P. falciparum* being the most lethal.

1.4.2 Plasmodium *spp.*

Plasmodium spp., imposition of fitness cost and malaria.

The major protozoan parasites treated in this section include *P. falciparum, P. malariae, P. ovale, P. vivax,* and *P. knowlesi*, the causative agents of human malaria. They undergo a complex life cycle involving liver, red blood cell stages in men, and sexual life stages in the female Anopheles mosquito vector. Pathophysiologically, parasites invade red blood cells, feed on the hemoglobin in these cells to derive nourishment, and rupture the cells leading to anemia. They also alter the surface of infected cells through the remolding of the cytoskeleton of the cells and exportation of antigens, including variant surface antigens, onto the cell surfaces, which facilitates adherence of the infected cells to the inner surfaces of blood vessels, thereby causing infected erythrocytes to sequester in the major organs like the brain and tissues in humans in general as well as in the placenta in the case of pregnancy-associated malaria as the parasites mature to the trophozoite and schizont stages, leading to complications such as cerebral malaria, respiratory distress, and spontaneous abortion among others [8].

Infection begins when an infected female Anopheles mosquito, injects sporozoites into the skin of a host. The sporozoites find their way into the liver through the bloodstream and traverse liver sinusoidal endothelial cells or Kupffer cells to infect hepatocytes [9–11]. Inside a hepatocyte, sporozoite divides into thousands of merozoites through a process of hepatocytic schizogony. Mature merozoites egress from the cell and enter the bloodstream only to invade erythrocytes and start a new cycle of schizogony within red blood cells (RBC), which includes asexual replication of their haploid genome. Within RBCs, parasites develop through a ring, trophozoite, and schizont stages. The resulting mature schizont is normally segmented in appearance and contains 16–32 daughter merozoites [12]. The infected RBC ruptures, releasing the daughter merozoites to invade new RBCs to start another cycle of development and asexual replication besides flooding the bloodstream with its by-products of parasite metabolism such as hemozoin formed when the parasite digests hemoglobin and cellular debris. This cycle of erythrocytic development takes approximately 24 hours for *P. knowlesi*, 48 hours for *P. falciparum, P. vivax*, and *P. ovale*, and 72 hours for *P. malarae*. The release of parasite materials into the bloodstream upon the rupture of the infected erythrocytes also triggers host responses leading to various clinical symptoms, including fever, chills, headache, dizziness, back pain, and myalgia. Some patients may progress to severe malaria with coma (cerebral malaria), pulmonary edema, acute renal failure, jaundice, severe anemia, acidosis, hypoglycemia, and death [8, 9].

When a mosquito takes a blood meal from an infected host, gametocytes are picked alongside the blood meal to undergo sexual development in the midgut of mosquitoes. Within a minute of entering the midgut of the mosquito, the male and female gametocytes transform into male and female gametes respectively. The male then fertilizes the female to form a zygote. The zygote develops into motile ookinetes, which traverse through the mosquito midgut wall to develop into oocysts containing several thousands of sporozoites, which then migrate to salivary glands upon maturation and are injected into a new vertebrate host when the mosquito bites again for a blood meal, thereby completing the life cycle [8, 9].

Clinically, the disease manifests as fever, chills, headache, nausea, and in severe cases, severe malaria anemia, organ failure, and death, the ultimate cost imposed on the host by the parasite. The infection is systemic but only in the ring stage of the parasites, and in some cases, the gametocytes are seen in the peripheral blood as the mature stages, and the mature trophozoites and schizonts sequester in the deeper tissues and organs [8, 9].

As the infection lingers on, *P.* ovale and *P. vivax,* transform into hypnozoites in hepatocytes and dormant only to cause blood stage infection after a couple of weeks to several years. In such a state the parasite does not seem to harm the host but benefits from the host, a symbiotic relationship. In this sense, the parasite derives free nourishment and shelter from the host and transport as the host moves from one place to another, including across continents to underpin its parasitic status while the host suffers, including the ultimate death (**Figure 5**) [8, 9].

The disease is a preventable and controllable one so vector control measures like in-door residual spraying of insecticide, destruction of the breeding places of mosquitoes, destruction of the mosquito larvae, prophylaxes and vaccine with RTS, S/AS01 being the first licensed vaccine in use with others under development are deployed to curtail its devastating effect on humans [12–14]. Following diagnosis based on clinical history, rapid diagnostic testing or light microscopy, which is the gold standard for diagnosis, it is treated with antimalarial drugs, including chloroquine, artemisinin-based combination therapies for the asexual or blood stage while the liver stage, including the hypnozoites as is the case *P. vivax* and *P. ovale*, is treated with primaquine [8].

1.4.3 Toxoplasma gondii

Toxoplasma gondii: This organism is responsible for toxoplasmosis, which poses serious problems to pregnant women and immunocompromised individuals including people living with HIV/AIDS. Infection is typically acquired through ingestion of oocysts from contaminated food or water, or contact with the feces of infected cats, the definitive hosts.

1.4.4 Cryptosporidium *spp.*

Cryptosporidium spp.: These parasites cause cryptosporidiosis, a gastrointestinal illness characterized by watery diarrhea. Transmission occurs through ingestion of oocysts in contaminated water or food. It is a major cause of diarrheal disease, particularly affecting children and immuno-compromised individuals.

1.4.5 The zoonotic parasite, Babesia *spp.*

The zoonotic parasite, *Babesia* spp., is responsible for babesiosis, a tick-transmitted disease that infects red blood cells of both humans and rodents without the red cells becoming large but vacuolated leading to symptoms such as fever, fatigue, and hemolytic anemia [1].

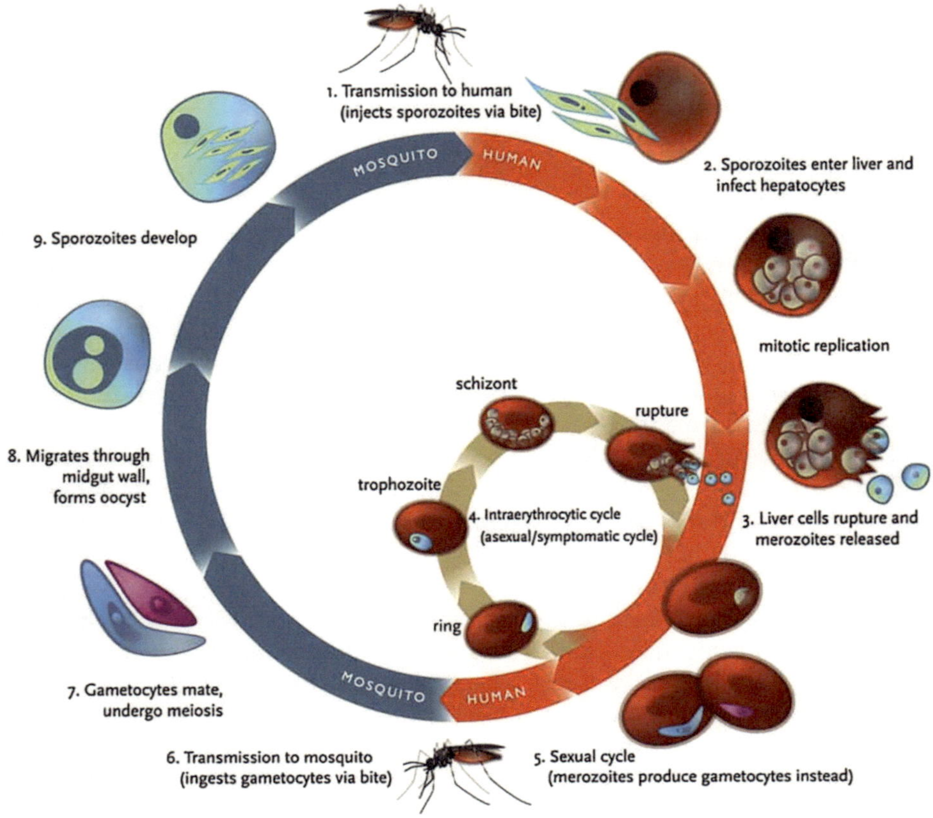

Figure 5.
Life cycle of malaria parasite. (1) In the human infective stage of the parasite, sporozoites are injected into a human during the subsequent blood meal or meals. (2) Sporozoites quickly travel to the liver, infect hepatocytes, and start replicating asexually or mitotically. (3) After roughly 6–15 days, the liver schizonts rupture, releasing thousands of merozoites into the bloodstream. (4) Merozoites enter red blood cells and begin the cyclic blood stage of infection. The parasites replicate and mature to the schizont stage, and the infected RBCs rupture to release more merozoite to infect new RBCs repeatedly. (5) Some of the parasites transform into sexual forms, male and female gametocytes. (6) Gametocytes are consumed by mosquitoes when they feed on an infected human. (7) Male and female gametocytes mate in the gut of the mosquito. (8) Oocyst develops, and undergoes meiosis, producing sporozoites in the process. (9) Sporozoites lodge in the salivary gland to be injected into another human during a blood meal (modified from https://onehealthtrust.org/publications/infographics/life-cycle-malaria-parasite/ on 2nd December 2024).

2. *The helminths*

Helminths, commonly known as parasitic worms, are among the most significant parasitic organisms infecting humans. They are multicellular eukaryotic organisms and have coexisted with humans for millennia, establishing with their hosts a complex relationship. A multifaceted relationship that impacts human health in various ways, causing both morbidity and mortality, particularly in areas with poor sanitation and hygiene. Helminths are broadly classified into three main taxonomical groups based on their morphology and biological features that contribute to their ability to infect and persist in the host. They include nematodes or roundworms, cestodes or tapeworms, and trematodes or flukes.

Cestodes are *Taenia solium* or the pork tapeworm and *Echinococcus granulosus*, which causes hydatid disease with trematodes being *Schistosoma* spp. Each group contains a variety of species that have evolved to infect different tissues and systems in their hosts. Major groups of helminths, their life cycles, pathogenicity, and effects on human health are discussed.

2.1 Nematodes (roundworms)

Nematodes (Roundworms): These organisms are the most diverse group of helminths found in various environments, including soil, and water, with some being parasitic while others are not. They are characterized by cylindrical, non-segmented bodies and a complete digestive system with separate openings for ingestion and excretion. Nematodes include *Ascaris lumbricoides*, or the giant roundworm, *Ancylostoma duodenale*, or the hookworm, and *Enterobius vermicularis*, or pinworm.

2.1.1 Ascaris lumbricoides

Ascaris lumbricoides: This parasitic worm, a soil-transmitted helminth, causes one of the most common helminth infections, Ascaris. The adult worms reside in the intestines, suck blood, and damage the intestinal wall thereby causing malnutrition, abdominal pain, and, in severe cases, intestinal obstruction. Ascaris infections are transmitted *via* the fecal-oral route, with eggs excreted in feces and ingested by humans through contaminated food or water while *Enterobius vermicularis* or pinworm, transmitted like Ascaris, causes infection that leads to even autoinfection (**Figure 6**) [15].

2.2 Cestode (tapeworms)

Cestode (Tapeworms). Cestodes exhibit intricate lifecycles involving intermediate and definitive hosts. In the case of *Taenia solium*, humans can function as both definitive and intermediate hosts. Consuming undercooked meat containing larval cysts of *Taenia solium* or pork tapeworm and *Taenia saginata* or beef tapeworm leads to infection with the larvae maturing into adult tapeworms in the intestines. Cysticercosis, a condition associated with the migration of larvae to other tissues, including the brain, leading to neurological symptoms and potentially life-threatening complications, occurs when humans inadvertently ingest eggs of the tapeworm from contaminated food or water that hatch into wandering larvae later.

2.2.1 Echinococcus granulosus *and* Echinococcus multilocularis

Echinococcus granulosus and *Echinococcus multilocularis* are associated with echinococcosis, characterized by hydatid cysts in organs such as the liver, lungs, and brain. This condition is contracted by consuming eggs expelled by infected canines, which serve as the definitive hosts. Consumption of undercooked fish carrying the larval form of *Diphyllobothrium latum*, a fish tapeworm, on the other hand, leads to diphyllobothriasis, an infection that can lead to vitamin B12 deficiency, to result in megaloblastic anemia. The *Echinococcus granulosus* life cycle involves canines (domestic or wild) as definitive hosts and herbivores (like sheep) as intermediate hosts, with humans becoming accidental intermediate hosts upon ingesting eggs [15].

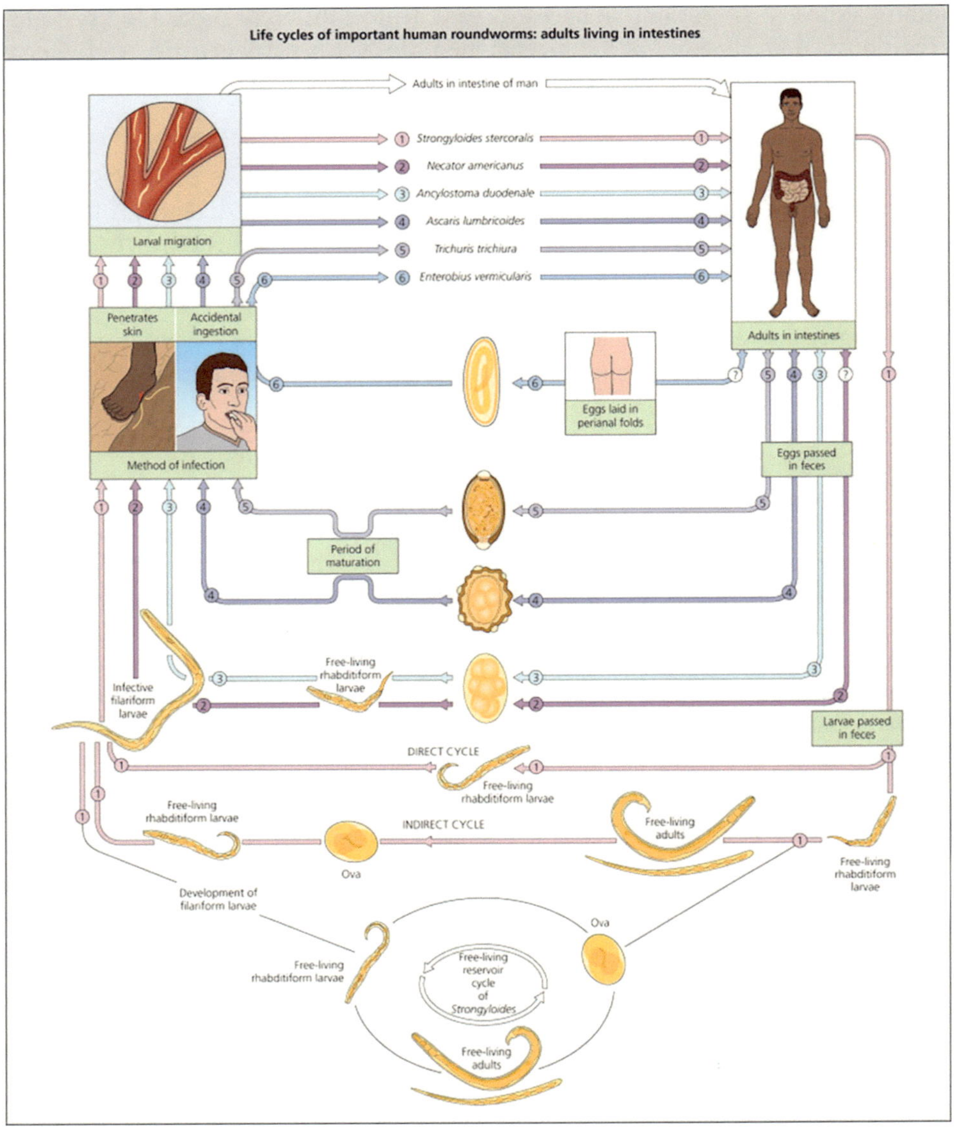

Figure 6.
Life cycles of human roundworms: Adults living stage living the intestine. The figure sumarizes the life cycle of Enterobius vermicularis, Ascaris lumbricoides, Strongyloides stercoralis, Necator americanus, Trichuris trichiura, Ancylostoma duodenale. *It shows the routes of infection, the human infective stage, and development in both humans and the environment where possible [15].*

2.3 *Trematodes (flukes)*

Trematodes, known as flukes, are leaf-shaped flatworms that infect various human tissues, including the liver, lungs, and blood vessels. These parasites are hermaphroditic, possessing both male and female reproductive organs, except for *Schistosoma* spp., which are sexually dimorphic with distinct sexes [16]. Trematode infections are particularly prevalent in regions with inadequate sanitation and limited access to potable or clean water (**Figure** 7).

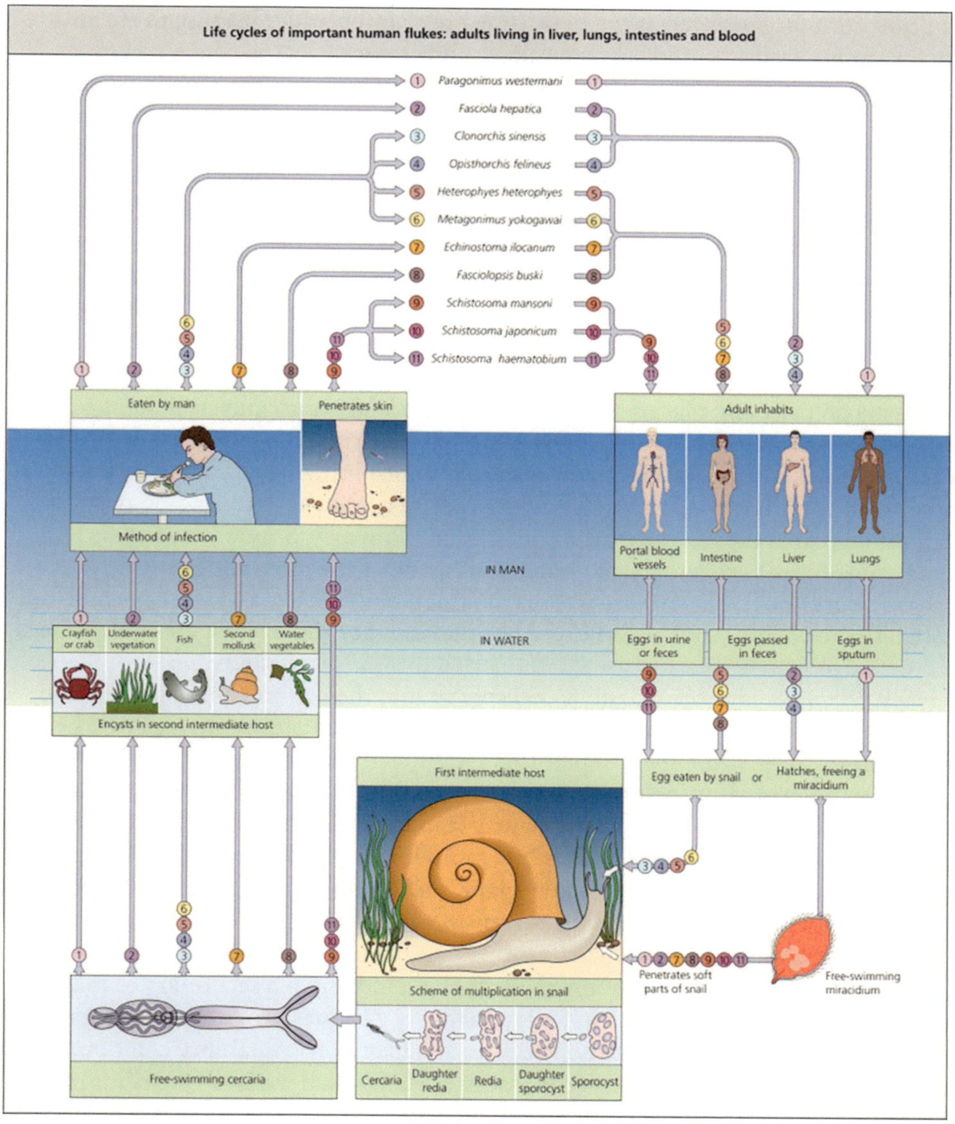

Figure 7.
Life cycles of trematodes. Adults living stage living the intestine. The figure summarizes the life cycle of Fasciola hepatica, Schistosoma *spp. Routes of infection, the infective stage to humans, and development in both humans and in the environment where possible are traced by the figure [15].*

2.3.1 The Schistosoma *spp*

The *Schistosoma* spp. These are responsible for schistosomiasis or bilharziasis, one of the most widespread parasitic diseases worldwide. The mature worms inhabit blood vessels, and the eggs released by females may enter the intestine and be voided out of the host together with feces in the case of intestinal schistosomiasis caused by *Schistosoma mansoni*, *S. guineensis, S. intercalatum*. *S. japonicum*, and *S. mekongi* or into the genitourinary system and be passed out in the urine in the case of genitourinary schistosomiasis caused by *S. hermatobium* [17]. The eggs also trigger inflammation and fibrosis in organs such as the liver, intestines, and bladder leading to encapsulation by immune and subsequent granuloma formation.

The transmission of helminths to humans typically occurs through ingestion of eggs and larvae in contaminated food or drinking water, skin penetration by larvae, or inoculation by insect vectors. Once in the human body, helminths establish themselves in various tissues and organs, including the gastrointestinal tract, lungs, liver, and bloodstream. The pathophysiology of infection depends on the ability of the parasite to evade the immune response of the host and persist for extended periods to cause mechanical, inflammatory, and immunological damage. The disease in humans occurs through direct tissue damage, competition for nutrients, and immune evasion. These mechanisms vary depending on the species of helminth and the stage of infection.

Tissue damage and obstruction: *Ascaris lumbricoides* cause physical damage as they migrate through tissues. Ascaris larvae can penetrate the intestinal wall, and reach the lungs *via* the bloodstream, resulting in respiratory symptoms. In heavy infections, adult Ascaris worms can cause intestinal obstruction, leading to severe abdominal pain, emesis, and bowel perforation. Similarly, *Schistosoma* spp. is responsible for schistosomiasis or bilharziasis, one of the most widespread parasitic diseases worldwide. The mature worms inhabit blood vessels, and the eggs released by females may enter the intestine and be voided out of the host together with feces in the case of intestinal schistosomiasis caused by *Schistosoma mansoni*, *S. guineensis*, *S. intercalatum*. *S. japonicum*, and *S. mekongi* or into the genitourinary system and be passed out in the urine in the case of genitourinary schistosomiasis caused by *S. hermatobium* [17]. *Schistosoma* spp. inflicts significant tissue damage as their eggs lodge in various organs, including the liver, intestines, bladder, and fallopian tube [18, 19]. Blockage of the fallopian tube may lead to infertility in women. The immune response to these eggs is associated with granuloma formation and fibrosis that results in damage or dysfunction of affected organs. For instance, in chronic schistosomiasis, hepatic fibrosis can lead to portal hypertension and severe gastrointestinal bleeding. Common occurrences during schistosome infection include a heightened risk of placental inflammation, nutrient malabsorption, and anemia secondary to maternal and fetal iron deficiency [20]. Pro-inflammatory cytokines production in response to schistosomes, which upregulate hepcidin and reduce iron availability compounds iron loss due to eggs shedding through the bladder or intestinal walls to contribute further to anemia [21]. Other helminths, like hookworms adhere to the intestinal lining and directly consume the blood of the host [22]. This deprives the host of nutrients and causes deficiency such as iron deficiency with consequential anemia.

3. Coevolution of parasites and humans

Parasites and humans have been interacting for years which has significantly shaped human genetics, immunity, and population dynamics. Plasmodium parasites have exerted powerful selective force on humans, particularly in regions where the disease is prevalent. This evolutionary arms race between humans, Plasmodium, and other pathogens typifies how these pathogens influence the development of host traits and vice versa.

At the genetic level, the high mortality rate of malaria, especially in children, has driven the selection of genetic traits that confer resistance to the disease. For instance,

a. Sickle cell trait: Individuals heterozygous for the sickle cell gene, hemoglobin S and C, are less susceptible to severe malaria, as the altered red blood cells are less friendly to Plasmodium [23–26].

b. Glucose-6-phosphate dehydrogenase (G6PD) deficiency impairs the replication within red blood cells due to increased oxidative stress [27].

c. Thalassemia: Mutations causing thalassemia have reduced the ability of the parasite to utilize red blood cells, providing a survival advantage in malaria-endemic areas [28].

3.1 At population level

Malaria has influenced human population structures by favoring genetic diversity in immune-related genes. Polymorphisms in genes encoding human leukocyte antigen (HLA) molecules and cytokines like tumor necrosis factor-alpha (TNF-α), interleukin (IL)-10, triggering receptors expressed on myeloid cells, haptoglobin, complement molecules are associated with varying levels of resistance to malaria, demonstrating the role of the parasites in maintaining genetic variation [29–31]. Traits like anemia and iron sequestration during helminth infections might represent evolutionary strategies to limit parasite survival [32].

3.2 Evolution of the human immune system

The persistent risk of malaria and other parasites has shaped human immune responses leading to adaptations that balance immunity and tolerance. Adaptive immunity to the parasites involves multifaceted interactions between innate and adaptive systems, such as the activation of T cells and the production of antibodies that target the parasites and their antigens. Following repeated exposure in regions endemic to malaria, individuals develop premunition, a nonsterile immunity to malaria proving how malaria has shaped memory and tolerance mechanisms of the immune system [33]. Plasmodium parasites have therefore evolved strategies to evade human defenses, such as antigenic variation, as is seen in a host of the parasites, including *Plasmodium* spp., *Leishmania* spp., *Trypanosoma* spp., and immunosuppression as is the case with both protozoan and helminth [34–36]. This dynamic interplay highlights the dual and reciprocal nature of coevolution, where human adaptations are met with parasite countermeasures and vice versa [37].

4. Antigenic variation and parasite secretary products in immune response modulation and evasion

The long life span of parasites in their host undoubtedly shows that they interfere with and modulate immune response from the moment they infect to coexist with the host. They do not simply ward off immune attacks; instead, they influence and direct immune responses away from the modes that severely damage them to create niches that optimize successful feeding, reproduction, and transmission [38].

In regions endemic to malaria, people develop premunition, a nonsterile immunity to malaria after repeated exposure, demonstrating how malaria has shaped immune system memory and tolerance mechanisms [33]. Antigenic variation confers an escape route from the immune response for some of these parasites. Variable antigenic surface proteins encoded by multicopy gene families [34–36], among others, and polymorphic alleles in infected erythrocytes at different stages of the parasite help malaria parasites to evade host immune recognition and attack [13, 34–36].

Trypanosoma brucei, the causative agent of African trypanosomiasis, lives extracellularly in the bloodstream of humans and masterfully evades the immune response through antigenic variation to coexist with the host. To avoid destruction by antibodies, *T. brucei* regularly changes proteins on its surface, specifically its variant surface glycoproteins (VSGs), to a new one, which the immune system must develop adaptive immunity to be more effective against the parasite. The immune system initially mounts a robust immune response against the parasite, producing antibodies that recognize VSGs. However, before the immune system can eliminate the infection, *T. brucei* switches to a new VSG variant to avoid detection and attack. This process allows *T. brucei* to persist in the host for extended periods, leading to chronic infection characterized by waves of parasitemia and fluctuating fevers. The constant stimulation of the immune system by various antigens of *T. brucei* through antigenic variation can also lead to immune exhaustion, which will weaken the immune defenses of the host and promote the survival of the parasites. It also appears as if the immune response directed against the parasite is a repeated innate immune response to the same pathogen that is not very effective like the adaptive immune responses since the constant inundation of the immune system with novel antigens renders the mounted adaptive immune response and memory to the previous antigens of the same parasite ineffective. Antigenic variation is employed by other protozoans like Leishmania spp. to outwit the immune system.

Adult schistosomes adsorb complement system activation regulatory molecules, the decay accelerating factor (DAF), or cluster of differentiation (CD)55 from host erythrocytes into their tegument to effectively interact with C3 to interrupt activation of both classical and alternative pathways of complement system activation to protect themselves [39]. Trepanning orphan receptor of *S. haematobium* (Sh-TOR) binds to human complement C2 to interfere with the activation beyond complement protein C2 and Schistosome complement inhibitory protein-1 (SCIP-1) that functions like CD59 to disrupt membrane attack complex (MAC) formation on self-cells, is expressed to interact with C8 and C9 to inhibit MAC formation to protect against complement-mediated lysis of the parasite [40, 41].

The merozoite stage of the malaria parasite recruits C1 esterase inhibitor (C1-INH), a negative regulator of complement activation, to its surface when exposed to human circulation and utilizes it to inhibit C1s, mannan-binding lectin serine protease 1 (MASP1), and MASP2 [42] thereby disrupting classical and lectin pathways of complement system activation to accord the merozoite protection. Similarly, *E. granulosus* adsorbs factor H (FH), a negative regulator of the complement system that protects cells from being attacked and destroyed by the body's complement system, *via* its Myo-inositol hexakisphosphate (InsP6), which is a major component of the acellular laminated layer (LL) of the hydatid cyst wall, onto its surface to prevent deposition of C3b, an excellent opsonin, and component of C5 convertase, on the surface of the parasite [43] to interfere with C5 convertase formation to abrogate the complement system activation.

Resisting intracellular killing as well as employment of a deceptive mechanism, trojan horse, to passively enter an otherwise hostile environment without appropriate hostility from the cell, the macrophage, is employed by leishmania parasites. Following entry into the human host, neutrophils rapidly engulf promastigotes but fail to kill them, instead, they undergo apoptosis, attracting macrophages [44] to engulf them along with viable Leishmania parasites, which then take advantage of the intracellular niches in the macrophages for replication and multiplication. Maturation of the phagosome maturation to kill the parasite following engulfment is resisted [45]. Furthermore, macrophage inducible Nitric Oxide Synthase expression is downregulated, reducing nitric oxide levels required for parasite killing while simultaneously upregulating host arginase, diverting arginine metabolism from nitric oxide synthesis to polyamine production to support the growth of the parasite [46, 47].

These parasites also secrete bioactive molecules into circulation to modulate the immune response for their survival and transmission. *Schistosoma mansoni* and eggs-specific antigens such as lysophosphatidylserine, lacto-N-fucopentaose III, and double-stranded ribonucleic acid (RNA) interact with pattern recognition receptors as toll-like receptors (TLRs) to attenuate inflammation. Experimentally, omega-1 from *Schistosoma mansoni* demonstrates Th2 response and suppresses intestinal inflammation in experimental models. Experimentally, lacto-N-fucopentaose III induces dendritic cell maturation and Th2 response polarization through C-type lectin receptors while filarial nematodes produce glycoprotein, excretory-secretory products (ES)-62, released by *Acanthocheilonema viteae*, *Schisosoma* spp., and others target TLR4 on host immune cells [48].

5. Mutualism: A rare aspect of helminthiasis

In a mutualistic relationship, both organisms benefit. While rare, some aspects of helminthiasis could be considered mutualistic, particularly in populations where the immune-modulating effects of helminths seem to protect against certain autoimmune diseases or allergic conditions. Helminths are found in the lumen of the gut but can influence gut-associated lymphoid tissue (GALT) and systemic immunity, often through the induction of immunoregulatory mechanisms that shape disease outcomes beyond the site of infection. Interaction between helminths and pattern recognition receptors (PRRs) on epithelial cells and antigen-presenting cells (APCs) in GALT triggers a type 2 immune response, activation of Th2 cells, eosinophils, mast cells, and the production of such anti-inflammatory cytokines as IL-4, IL-5, and IL-13 [38], which suppresses the Th1 and Th17 responses involved in inflammation-associated disorders [39]. Expectedly, an inverse relationship has been observed between prevalences of inflammatory disorders such as inflammatory bowel disease (IBD), asthma, and helminths [49, 50]. Moreso, helminths and their secreted products promote polarization of regulatory T cells (Tregs), development of tolerogenic phenotypes in dendritic cells (DCs) and alternative activation of macrophages (M2 phenotype), which together promote the release of factors such as IL-10 and transforming growth factor β (TGF-β) to limit inflammation and enhances immune tolerance [22, 51–53] such that Trichuris suis therapy seems useful in treating such inflammatory disorders as inflammatory bowel disease (IBD), multiple sclerosis (MS) and Crohn's disease [54–55].

More so, the capacity to induce regulatory responses suppresses the effector phase of the allergic response [56]. This reaction is produced *via* proteins secreted by worms [47, 57–59] such as ES-62, and lacto-N-fucopentaose III (LPI). A high level of exposure to infectious agents in early childhood decreases the likelihood of developing an allergy later in life [60]. In societies with high exposure to pathogens, the immune system may become less overactive (**Figure 8**), resulting in lower rates of allergic disorders. In such societies, autoimmune diseases such as multiple sclerosis,

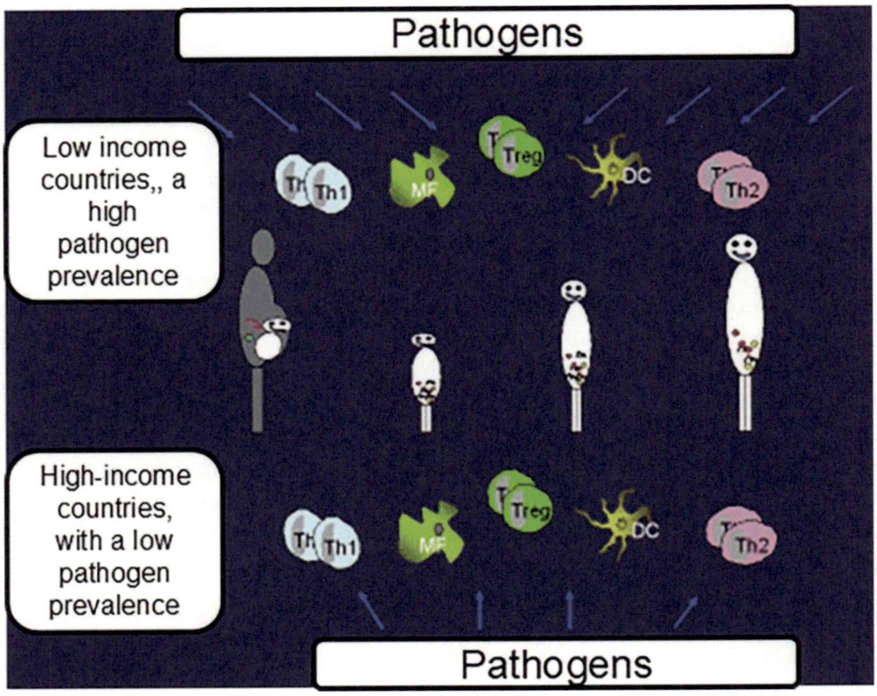

Figure 8.
Relationship between early and high exposure to pathogens and inflammatory disorders. People in societies with cleaner environments experience high inflammatory disorders later in life while the converse seems true for people exposed to a lot of pathogens early in life.

Crohn's disease, and type 1 diabetes are significantly less prevalent [61], suggesting that helminths might confer some benefit on their human hosts under certain conditions.

Symbiosis, therefore, seems to be a continuum of fitness outcomes, the cost incurred by a mutualist for its service and the virulence of a parasite may be how much resource the parasite takes from the host, fitness losses incurred by the host because of infection as it reduces the fitness of the host in the process, and in some cases killing the host. In the event of killing the host, both parasites and the host lose out since the parasite will not be transmitted to another host.

For humans, these adaptations are likely to reduce the risk of autoimmune diseases in environments where helminths have been endemic. In contrast, however, the decline in helminth exposure, particularly in modern and industrialized societies where the environments are cleaner, has been associated with increasing prevalence of autoimmune and allergic diseases support "hygiene hypothesis" [49].

The reduced interaction of humans and helminth exposure in contemporary life has disrupted the evolutionary equilibrium between these two, contributing to immune dysregulation and increased prevalence of inflammatory disorders. Unexpectedly, helminth-derived molecules are now being explored as therapeutic products for autoimmune and allergic diseases, leveraging their immunomodulatory properties [61].

In all this, however, the benefits do not outweigh the numerous negative consequences of helminth and protozoan infections, including malnutrition, anemia, and impaired cognitive development, particularly in children with the association being more parasitic and exploitative than mutualism. Therefore, while there may be mutualistic aspects in specific contexts, the overall relationship between helminths and protozoans and their human host remains exploitative on the part of the parasite. Science is, however, evolving, and there may be more insight into the mutualistic aspect of this association, either directly or indirectly soon.

Author details

Selorme Adukpo
Department of Pharmaceutics and Microbiology, School of Pharmacy, University of Ghana, Legon, Ghana

*Address all correspondence to: sadukpo@noguchi.ug.edu.gh

References

[1] Symth JD. Animal Parasitology. 3rd ed. Great Britain: Cambridge University Press; 1996. pp. 130-136

[2] Kelso AA, Waldvogel SM, Luthman AJ, Sehorn MG. Homologous Recombination in Protozoan Parasites and Recombinase Inhibitors. Frontiers in Microbiology. 2017;**8**:1716. DOI: 10.3389/fmicb.2017.01716

[3] Büscher P, Cecchi G, Jamonneau V, Priotto G. Human African trypanosomiasis. Lancet (London, England). 2017;**390**(10110):2397-2409. DOI: 10.1016/S0140-6736(17)31510-6

[4] Lejon V, Bentivoglio M, Franco JR. Human African trypanosomiasis. Handbook of Clinical Neurology. 2013;**114**:169-181. DOI: 10.1016/B978-0-444-53490-3.00011-X

[5] Hochberg NS, Montgomery SP. Chagas disease. Annals of Internal Medicine. 2023;**176**(2):ITC17–ITC32. DOI: 10.7326/AITC202302210

[6] Pérez-Molina JA, Molina I. Chagas disease. Lancet (London, England). 2018;**391**(10115):82-94. DOI: 10.1016/S0140-6736(17)31612-4

[7] Schuster FL, Ramirez-Avila L. Current world status of Balantidium coli. Clinical Microbiology Reviews. 2008;**21**(4):626-638. DOI: 10.1128/CMR.00021-08

[8] Jiang H, Li N, Gopalan V, Zilversmit MM, Varma S, Nagarajan V, et al. High recombination rates and hotspots in a plasmodium falciparum genetic cross. Genome Biology. 2011;**12**(4):R33. DOI: 10.1186/gb-2011-12-4-r33

[9] Trampuz A, Jereb M, Muzlovic I, et al. Clinical review: Severe malaria. Critical Care. 2003;7:315. DOI: 10.1186/cc2183

[10] Tavares J, Formaglio P, Thiberge S, Mordelet E, Van Rooijen N, Medvinsky A, et al. Role of host cell traversal by the malaria sporozoite during liver infection. The Journal of Experimental Medicine. 2013;**210**(5):905-915. DOI: 10.1084/jem.20121130

[11] Ménard R, Tavares J, Cockburn I, Markus M, Zavala F, Amino R. Looking under the skin: The first steps in malarial infection and immunity. Nature Reviews. Microbiology. 2013;**11**(10):701-712. DOI: 10.1038/nrmicro3111

[12] Su XZ, Zhang C, Joy DA. Host-malaria parasite interactions and impacts on mutual evolution. Frontiers in Cellular and Infection Microbiology. 2020;**10**:587933. DOI: 10.3389/fcimb.2020.587933

[13] Adukpo S. Malaria Vaccine Development and Associated Challenges. London, UK: IntechOpen; 2024. DOI: 10.5772/intechopen.115131

[14] Wadman M. First malaria vaccine slashes childhood deaths. Science (New York, N.Y.). 2023;**382**(6669):357. DOI: 10.1126/science.adl5521

[15] Alan Lindquist HD, John H. Cross, 195- Helminths, Infectious Diseases. 4th ed. Elsevier; 2017. pp. 1763-1779

[16] Marco A, Kozomara A, Hui JH, Emery AM, Rollinson D, Griffiths-Jones S, et al. Sex-biased expression of microRNAs in Schistosoma mansoni. PLoS Neglected Tropical Diseases.

2013;**7**(9):e2402. DOI: 10.1371/journal.pntd.0002402

[17] Kołodziej P, Szostakowska B, Lass A, Sulima M, Sikorska K, Kocki J, et al. Chronic intestinal schistosomiasis caused by co-infection with Schistosoma intercalatum and Schistosoma mansoni. The Lancet. Infectious Diseases. 2024;**24**(3):e196-e205. DOI: 10.1016/S1473-3099(23)00486-3

[18] Schroers B, Peschen-van Issum M, Reinert RR, Füzesi L, Biesterfeld, Winkler M. Sterilität als Folge einer tubaren Schistosomiasis [sterility as a sequela of tubal schistosomiasis]. Geburtshilfe und Frauenheilkunde. 1995;**55**(3):140-142. DOI: 10.1055/s-2007-1022792

[19] Ekoukou D, Luzolo-Lukanu A, Mulard C, Bazin C, Ng Wing Tin L. Bilharziose péritonéo-tubaire à Schistosoma haematobium. A propos de 2 cas [Peritoneal and tubal Schistosoma haematobium bilharziasis. Two case reports]. Journal de Gynecologie, Obstetrique et Biologie de la Reproduction. 1995;**24**(8):819-824

[20] Ajanga A, Lwambo NJ, Blair L, Nyandindi U, Fenwick A, Brooker S. Schistosoma mansoni in pregnancy and associations with anaemia in Northwest Tanzania. Transactions of the Royal Society of Tropical Medicine and Hygiene. 2006;**100**(1):59-63. DOI: 10.1016/j.trstmh.2005.06.024

[21] Ganz T. The role of hepcidin in iron sequestration during infections and in the pathogenesis of anemia of chronic disease. The Israel Medical Association Journal: IMAJ. 2002;**4**(11):1043-1045

[22] McSorley HJ, Maizels RM. Helminth infections and host immune regulation. Clinical Microbiology Reviews. 2012;**25**(4):585-608. DOI: 10.1128/CMR.05040-11

[23] Allison AC. Protection afforded by sickle-cell trait against subtertian malarial infection. BMJ. 1954;**1**(4857):290-294

[24] Weatherall DJ, Clegg JB. Inherited haemoglobin disorders: an increasing global health problem. Bulletin of the World Health Organization. 2001;**79**(8):704-712

[25] Mangano VD, Kabore Y, Bougouma EC, Verra F, Sepulveda N, Bisseye C, et al. Novel insights into the protective role of hemoglobin S and C against plasmodium falciparum parasitemia. The Journal of Infectious Diseases. 2015;**212**(4):626-634. DOI: 10.1093/infdis/jiv098

[26] Agarwal A, Guindo A, Cissoko Y, Taylor JG, Coulibaly D, Koné A, et al. Hemoglobin C associated with protection from severe malaria in the Dogon of Mali, a west African population with a low prevalence of hemoglobin S. Blood. 2000;**96**(7):2358-2363

[27] Kwiatkowski DP. How malaria has affected the human genome and what human genetics can teach us about malaria. American Journal of Human Genetics. 2005;**77**(2):171-192

[28] Bozdech Z, Ginsburg H. Antioxidant defense in Plasmodium falciparum--data mining of the transcriptome. Malaria Journal. 2004;**3**:23. DOI: 10.1186/1475-2875-3-23

[29] Adukpo S, Gyan BA, Ofori MF, Dodoo D, Velavan TP, Meyer CG. Triggering receptor expressed on myeloid cells 1 (TREM-1) and cytokine gene variants in complicated and uncomplicated malaria. Tropical Medicine and International Health:

TM and IH. 2016;**21**(12):1592-1601. DOI: 10.1111/tmi.12787

[30] Hedrick PW. Population genetics of malaria resistance in humans. Heredity. 2011;**107**(4):283-304

[31] Fumagalli M et al. Parasites represent a major selective force for interleukin genes in humans. Nature Genetics. 2011;**43**(12):1159-1165

[32] Allen JE, Sutherland TE. Host protective roles of type 2 immunity: Parasite killing and tissue repair. Current Opinion in Immunology. 2014;**30**:93-99

[33] Langhorne J, Ndungu FM, Sponaas AM, Marsh K. Immunity to malaria: more questions than answers. Nature Immunology. 2008;**9**(7):725-732. DOI: 10.1038/ni.f.205

[34] Wahlgren M, Goel S, Akhouri RR. Variant surface antigens of plasmodium falciparum and their roles in severe malaria. Nature Reviews. Microbiology. 2017;**15**(8):479-491. DOI: 10.1038/nrmicro.2017.47,16

[35] Belachew EB. Immune response and evasion mechanisms of plasmodium falciparum parasites. Journal of Immunology Research. 2018;**2018**:6529681. DOI: 10.1155/2018/6529681

[36] Mwakalinga SB, Wang CW, Bengtsson DC, Turner L, Dinko B, Lusingu JP, et al. Expression of a type B RIFIN in plasmodium falciparum merozoites and gametes. Malaria Journal. 2012;**11**:429. DOI: 10.1186/1475-2875-11-429

[37] Mackinnon MJ, Marsh K. The selection landscape of malaria parasites. Science. 2010;**328**(5980):866-871

[38] Allen JE, Maizels RM. Diversity and dialogue in immunity to helminths. Nature Reviews. Immunology. 2011;**11**(6):375-388. DOI: 10.1038/nri2992

[39] Maizels RM, Balic A, Gomez-Escobar N, Nair M, Taylor MD, Allen JE. Helminth parasites--masters of regulation. Immunological Reviews. 2004;**201**:89-116. DOI: 10.1111/j.0105-2896.2004.00191.x

[40] Shao S, Sun X, Chen Y, Zhan B, Zhu X. Complement evasion: An effective strategy that parasites utilize to survive in the host. Frontiers in Microbiology. 2019;**10**:532. DOI: 10.3389/fmicb.2019.00532

[41] Parizade M, Arnon R, Lachmann PJ, Fishelson Z. Functional and antigenic similarities between a 94-kD protein of Schistosoma mansoni (SCIP-1) and human CD59. The Journal of Experimental Medicine. 1994;**179**(5):1625-1636. DOI: 10.1084/jem.179.5.1625), 10.1084/jem.179.5.1625)

[42] Deng J, Gold D, LoVerde PT, Fishelson Z. Mapping of the complement C9 binding domain in paramyosin of the blood fluke Schistosoma mansoni. International Journal for Parasitology. 2007;**37**(1):67-75. DOI: 10.1016/j.ijpara.2006.09.011

[43] Kennedy AT, Wijeyewickrema LC, Huglo A, Lin C, Pike R, Cowman AF, et al. Recruitment of human C1 esterase inhibitor controls complement activation on blood stage plasmodium falciparum merozoites. Journal of immunology (Baltimore, Md.: 1950). 2017;**198**(12):4728-4737. DOI: 10.4049/jimmunol.1700067

[44] Irigoín F, Laich A, Ferreira AM, Fernández C, Sim RB, Díaz A. Resistance of the Echinococcus granulosus cyst wall to complement activation: Analysis of the role of InsP6 deposits. Parasite

Immunology. 2008;**30**(6-7):354-364. DOI: 10.1111/j.1365-3024.2008.01034.x

[45] van Zandbergen G, Klinger M, Mueller A, Dannenberg S, Gebert A, Solbach W, et al. Cutting edge: Neutrophil granulocyte serves as a vector for leishmania entry into macrophages. Journal of Immunology (Baltimore, Md. : 1950). 2004;**173**(11):6521-6525. DOI: 10.4049/jimmunol.173.11.6521

[46] Matheoud D, Moradin N, Bellemare-Pelletier A, Shio MT, Hong WJ, Olivier M, et al. Leishmania evades host immunity by inhibiting antigen cross-presentation through direct cleavage of the SNARE VAMP8. Cell Host and Microbe. 2013;**14**(1):15-25. DOI: 10.1016/j.chom.2013.06.003

[47] Calegari-Silva TC, Vivarini ÁC, Pereira RMS, Dias-Teixeira KL, Rath CT, Pacheco ASS, et al. Leishmania amazonensis downregulates macrophage iNOS expression via histone deacetylase 1 (HDAC1): A novel parasite evasion mechanism. European Journal of Immunology. 2018;**48**(7):1188-1198. DOI: 10.1002/eji.201747257

[48] Zakeri A, Whitehead BJ, Stensballe A, de Korne C, Williams AR, Everts B, et al. Parasite worm antigens instruct macrophages to release immunoregulatory extracellular vesicles. Journal of Extracellular Vesicles. 2021;**10**(10):e12131. DOI: 10.1002/jev2.12131

[49] Strachan DP. Hay fever, hygiene, and household size. BMJ. 1989;**299**(6710):1259-1260. DOI: 10.1136/bmj.299.6710.1259

[50] Elliott DE, Summers RW, Weinstock JV. Helminths as governors of immune-mediated inflammation. International Journal for Parasitology. 2007;**37**(5):457-464. DOI: 10.1016/j.ijpara.2006.12.009

[51] McSorley HJ et al. Immunomodulation by helminth parasites: Defining mechanisms and mediators. International Journal for Parasitology. 2012;**42**(3):301-3100

[52] McSorley HJ, Loukas A, Pearson MS. Helminths in the hygiene hypothesis: Soften the debate. Trends in Parasitology. 2013;**29**(3):150-157

[53] Finlay CM, Walsh KP, Mills KH. Induction of regulatory cells by helminth parasites: Exploitation for the treatment of inflammatory diseases. Immunological Reviews. 2014;**259**(1):206-230

[54] Summers RW et al. Trichuris suis therapy in Crohn's disease. Gut. 2005;**54**(1):87-90

[55] Strachan DP. Hay fever, hygiene, and household size. BMJ. 1989;**299**(6710):1259-1260

[56] Yazdanbakhsh M, Kremsner PG, van Ree R. Allergy, parasites, and the hygiene hypothesis. Science (New York, N.Y.). 2002;**296**(5567):490-494. DOI: 10.1126/science.296.5567.490

[57] Caraballo L, Zakzuk J, Acevedo N. Helminth-derived cystatins: the immunomodulatory properties of an Ascaris lumbricoides cystatin. Parasitology. 2021:1-13. DOI: 10.1017/S0031182021000214

[58] Joardar N, Mondal C, Sinha Babu SP. A review on the interactions between dendritic cells, filarial parasite and parasite-derived molecules in regulating the host immune responses. Scandinavian Journal of Immunology. 2021;**93**(4):e13001. DOI: 10.1111/sji.13001

[59] DP. Hay fever, hygiene, and household size. BMJ (Clinical Research ed.). 1989;**299**(6710):1259-1260. DOI: 10.1136/bmj.299.6710.1259 12675754

[60] Pantazi AC, Mihai CM, Balasa AL, Chisnoiu T, Lupu A, Frecus CE, et al. Relationship between gut microbiota and allergies in children: A literature review. Nutrients. 2023;**15**(11):2529. DOI: 10.3390/nu15112529

[61] Murdaca G, Greco M, Borro M, Gangemi S. Hygiene hypothesis and autoimmune diseases: A narrative review of clinical evidences and mechanisms. Autoimmunity Reviews. 2021;**20**(7):102845. DOI: 10.1016/j.autrev.2021.102845